Data Science Using Oracle Data Miner and Oracle R Enterprise

Transform Your Business Systems into an Analytical Powerhouse

Sibanjan Das

Apress®

Data Science Using Oracle Data Miner and Oracle R Enterprise

Sibanjan Das
Pune, Maharashtra, India

ISBN-13 (pbk): 978-1-4842-2613-1   ISBN-13 (electronic): 978-1-4842-2614-8
10.1007/978-1-4842-2614-8

Distributed to the book trade worldwide by Springer Science+Business Media New York, 233 Spring Street, 6th Floor, New York, NY 10013. Phone 1-800-SPRINGER, fax (201) 348-4505, e-mail orders-ny@springer-sbm.com, or visit www.springeronline.com. Apress Media, LLC is a California LLC and the sole member (owner) is Springer Science + Business Media Finance Inc (SSBM Finance Inc). SSBM Finance Inc is a **Delaware** corporation.

For information on translations, please e-mail rights@apress.com, or visit www.apress.com.

Apress and friends of ED books may be purchased in bulk for academic, corporate, or promotional use. eBook versions and licenses are also available for most titles. For more information, reference our Special Bulk Sales–eBook Licensing web page at www.apress.com/bulk-sales.

Any source code or other supplementary materials referenced by the author in this text are available to readers at www.apress.com. For detailed information about how to locate your book's source code, go to www.apress.com/source-code/. Readers can also access source code at SpringerLink in the Supplementary Material section for each chapter.

Printed on acid-free paper

I dedicate this book

*To my parents, grandparents, sisters, brothers, and brothers-in-law
who motivated me to do what I always wanted to do*

To my Wife and six-month old son Sachit for their love and care

*To my mentors and Coaches Karthik Krishna Jonnalagadha,
John Kopchak, Prabir Sen, Bibhudatta Ghadei, Harishchandra Rakh, and
Nirmal Babu who has enriched my life with guidance and lessons
that help me to be a better technocrat and person every day.*

Contents at a Glance

Contents

About the Author

Sibanjan Das is a Business Analytics and Data Science consultant. He has over six years of experience in the IT industry working on enterprise resource planning (ERP) systems, implementing predictive analytics solutions in business systems and Internet of Things (IoT). An enthusiastic and passionate professional about technology and innovation, he has had the passion for wrangling with data from the early days of his career. He also enjoys reading, writing, and networking. His writings have appeared in various Analytics Magazines, and Klout (http://klout.com) has rated him among the top 2% professionals in the world talking about Artificial Intelligence, Machine Learning, Data Science, and IoT.

Sibanjan holds a Masters in IT degree with a major in Business Analytics from Singapore Management University, Singapore, and is a Computer Science Engineering graduate from Institute of Technical Education and Research, India. He is a Six Sigma Green Belt from the Institute of Industrial Engineers and also holds several industry certifications such as OCA, OCP, CSCMS, and ITIL V3.

About the Technical Reviewer

Ajit Jaokar is the Director of the AI/Deep Learning labs for Future cities at UPM (Universidad Politécnica de Madrid). Ajit's work spans research, entrepreneurship, and academia relating to IoT, predictive analytics, and mobility. His current research focus is on applying data science algorithms to IoT applications. This research underpins his teaching at Oxford University (Data Science for IoT) and the "City Sciences" program at UPM. Ajit lives in London and is a British citizen.

He has been nominated to the World Economic Forum's "Future of the Internet" council, board of Connected Liverpool (Smart city), and World Smart Capital program (Amsterdam). Ajit moderates/chairs Oxford University's Next generation mobile applications panel. Ajit has been involved in IoT-based roles for the webinos project (Seventh Framework Program [Fp7]) and has spoken at MobileWorld Congress, CTIA (The Wireless Association), CEBIT, Web20 expo, European Parliament, Stanford University, MIT Sloan, Fraunhofer Institute for Open Communication Systems (FOKUS), and the University of St. Gallen. He has been involved in transatlantic technology policy discussions. Ajit is passionate about teaching Data Science to young people through Space Exploration working with Ardusat.

About the Technical Reviewer

Acknowledgments

I wish to express my gratitude to each and every one who has contributed to my career and success. This is my first attempt at writing a book. I am equally excited and wish that everyone who reads this book finds it worthy. The list of people who influence me is long. Keeping the length of text in mind, I would like to thank a few of them who have made a lot of difference to me and in my career.

I am grateful to my colleagues and friends in Oracle and Zensar for their encouragement and support. Thank you Abhijit, Pravin, Dinesh, Sandhya, Anand, Sanjay, Eulalio, Sanjiv, Rajesh G., Team Toshiba, and Team Co-Development for believing in me and providing me an opportunity to work with you all.

Personal thanks to Pradeep, Sangeetha, Karthik, Charlie, Boriana, Ari, Vinayak, Mohan, Vijay, Puskala, Swathi, Rajesh, Naresh, and VoF Team for their helping hand and constant support for my work during my tenure at Oracle. I learned a lot from you all. The experience gained from working alongside you has provided the grist for this book.

Huge thanks to my friends and professors at Singapore Management University—Prof. Jing Jiang, Dr. Kam Tin Seong, Prof. Prabir Sen, Venkata Narayanan, Koo Ping Shung, and Murphy Choy—for sharing their knowledge and experience with me. Thank you Ankit, Mudit, Ai Bowen, Mugdha ,Edwin, Shantanu, Upkar, and Maruthi for being my partners in projects and crime.

Many thanks to my friends Sarat, Itishree, Vikas Kumar, Biswajit, Hamid, and Trojan Horses for your constant support, sharing drinks, and lending your ears to me when I needed it most.

Final thanks to the Apress team who worked behind the scenes to create a book out of my drafts of thoughts. Thank you, Suresh John Celestin, for reaching out to me to write the book. The book was made better by Apress Editors Prachi Metha, Ajit Jaokar, and Poonam Jain. Your comments and edits were valuable for the book.

Introduction

We are now in an era where consumer preferences are changing at lightning speed, social media is making the reviews transparent, and new competitors are appearing out of nowhere. Digitalization is starting to rule the world, and employee attrition has increased. To surf these new waves of business challenges, organizations are getting driven by technologies and deploying a variety of new enterprise applications like ERP Systems, CRM (customer relationship management) tools, Big Data, and Cloud-based solutions. Also, organizations are relying on data science technologies and algorithms to be intelligent and proactive. Data science, which is a relatively new term in the Business and IT industry, has hit the main stream for investment in the last few years. This is because it provides organizations the opportunity to gain insights about their business and drive proactive planning. Although getting some information or insights out of the data has been there since a decade ago, proactive planning is relatively new. Organizations are now able to predict what might happen in the future, which helps them to formulate strategies and plan in advance for the possible roadblocks. Statistical models and quantitative analysis sits at the core of data science and measures past performance to predict the future.

But this is not the end of the story. To keep up with everyone's expectations, organizations need to attend the customers as soon as they need service, retain employees before they resign, or identify frauds before they happen. They also have to be operationally efficient and at the same time cost-effective. This requires the ability to apply analytical insights to business operations at the right time. The only way to stay ahead of everyone is to be agile at data-driven, decision-making capabilities. This is possible when algorithms are brought near to the origin of data such as databases and data science routines are automated and integrated to business process workflows.

Data science involves a wide array of technologies and statistical algorithms. This makes it difficult to automate each and every aspect of it. However, there are some areas in data science that can be automated using scripts and workflows.

At a high level, we can classify data science automation into the following categories:

Repetitive tasks automation: Repetitive tasks are those that have to be done every time while building models. Data extraction, data cleaning, and basic data transformations such as imputing null values and algorithm-specific transformations are some tasks that fall into this category. These are to be done even for the same set of data every time. Automating these tasks would take some of the burdens off of a data scientist so they can concentrate more on solving business problems.

Automated statistician: This is an area of data science automation where statistical routines and machine learning are automated. The system executes the best algorithm based on the provided data set. It hides the intricacies and mathematical complexity of algorithms from the user. The user needs to provide automated statisticians with data.

It understands the data, creates different mathematical models, and returns the result based on the model that best explains the data. It is still at a nascent stage and an active area of research.

Problem-specific automation: This involves automating a data science process based on the problem at hand. This eases a user of carrying out the same activity for a specific problem several times. For example, the data scientist of an organization develops a model for predicting future sales of an organization. Based on an organization's requirement, this activity has to be carried out at different time frames, say monthly or quarterly. Each time, a wide array of tasks have to be performed, such as extracting the data, preprocessing it, creating statistical models, and returning the results back to the business operation's database. If this whole process is automated, it drives agile business decisions, better customer focus, and operationally efficient and managed resources.

The complexity of automation rises when we move up the pyramid of data science automation techniques as shown in Figure 1. An automated statistician sits at the top of complexity, as this requires the system to learn the input data patterns, find the best fit values, and self-optimize its parameters using several statistical and machine learning algorithms. This requires a generalization of various algorithm constraints and huge computing power. Problem-specific and repetitive-tasks automation are relatively easier to implement.

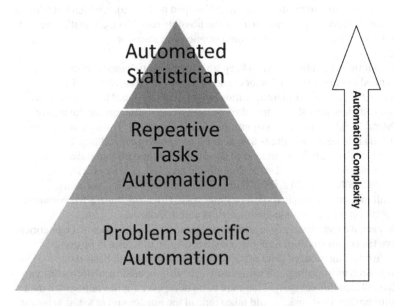

Figure 1. Data science automation pyramid

Oracle Advanced Analytics: Bringing Automation to Data Science

Oracle Advanced Analytics, an analytics product from Oracle, helps to address problem-specific automation and repetitive-tasks automation. It's got features like automated data preparation and workflow that integrate data science modules to business operations databases. It also has a feature known as predictive queries, one of the forms of automated statistics, for professionals with no statistics knowledge to be able to perform predictive analytics.

This book will help you learn about Oracle Advanced Analytics, focus on database embedded automation workflows for business use cases, and provide an overview on the most commonly used data science techniques for business applications. Rather than covering every detail of various machine learning algorithms, this book aims to give you a head start to drive automation and implement data science using Oracle Advanced Analytics. It also contains solutions to some practical data science problems that organizations need to embed to their business processes.

This book contains 8 chapters. Every chapter has various examples to demonstrate the capabilities of Oracle Advanced Analytics. In each chapter, I will outline some different statistical techniques, methods for automating different steps and implementing solutions. Toward the end of the book, you will have a good grasp and be able to perform in database automated analytics using Oracle Advanced Analytics technology stack.

The following is an overview of the contents for each chapter.

Chapter 1: Getting Started with Oracle Advanced Analytics

In this chapter, I will set the scene for you to know about data science, machine learning, and Oracle Advanced Analytics. I will introduce you to Oracle Data Miner and Oracle R Enterprise, which are a part of Oracle Advanced Analytics along with various PLSQL and ORE APIs that they offer for data mining and machine learning.

Chapter 2: Installation and Hello World

In this chapter, I will outline the steps to install Oracle Data Miner and Oracle R enterprise. I willalso demonstrate a "Hello World" workflow in SQL Developer to verify installation and get started with Oracle Advanced Analytics.

Chapter 3: Clustering Methods

In this chapter, you learn about different clustering methods available in Oracle Advanced Analytics stack. I will demonstrate an embedded workflow for RFM analysis and customer segmentation using some of the algorithms discussed.

Chapter 4: Association Rules

In this chapter, I discuss association rule mining in detail, followed by an example of market basket analysis embedded in an Organization's order management module to drive product recommendations.

Chapter 5: Regression Analysis

In this chapter, I brief you about regression techniques and their applications. This is followed by an example on predicting future sales for a superstore.

Chapter 6: Classification Methods

This chapter will help you learn about various classification options available in the Oracle Advanced Analytics stack. I demonstrate a classic example of predicting customer churn for an organization.

Chapter 7: Advanced Topics

In this chapter, I will brief you about ensemble methods, neural networks, and anomaly detection using Oracle Advanced Analytics.

Chapter 8: Solution Deployment

In this chapter, I demonstrate various methods to select the best model and deploy it to a production environment

CHAPTER 1

■ ■ ■

Getting Started With Oracle Advanced Analytics

In past few years, there has been a tremendous growth in unstructured as well as structured data, and they continue growing exponentially. Organizations are looking out for ways to derive value and make sense out of this data. They are looking for answers from the data that they have never asked before. To derive insights and make predictions from this data, they are using various statistical analytics tools such as R, Python, or SAS (Statistical Analysis System). However, integrating these analytical tools with business operation systems has some bottlenecks:

- There is a delay involved in moving data between systems.

- There is dependency on multiple resources and technologies to support and monitor this integration.

- Data quality has to be checked at every integration point.

- Integrating different systems costs time and money.

- Implementation of machine learning algorithms are resource intensive; and when the dataset is huge, it becomes more complicated.

- Open source tools like R are not very scalable when it comes to Oracle Enterprise implementations and cannot handle Big Data

This results in increasing costs and impacts the business. Since more than a decade ago, oracle database is recognized as one of the most popular, trusted, and widely used database systems by organizations. It's superior performance and security features has made it a reliable database system for managing large business operations. Oracle Advanced Analytics extends oracle database capabilities into an analytical platform

---

Electronic supplementary material The online version of this chapter (doi:10.1007/978-1-4842-2614-8_1) contains supplementary material, which is available to authorized users.

S. Das, *Data Science Using Oracle Data Miner and Oracle R Enterprise*, DOI 10.1007/978-1-4842-2614-8_1

to deliver insights from data. It leverages the strength of the database for parallelism, scalability, and security along with proprietary in-database implementations of commonly used predictive analytics machine learning techniques. It runs at the top of the oracle database and so does not need integration with the database.

In this chapter, I cover the following topics:

- Overview of Data Science and CRISP-DM (Cross-Industry Standard Process for Data Mining) Methodology

- Overview of machine learning and its application in various Industries

- Getting started with Oracle Advanced Analytics- Oracle Data Miner and Oracle R Enterprise

- Overview of some Analytical SQL (structured query language) and PLSQL (procedural language/SQL) functions that makes Oracle Advanced Analytics a powerful analytical engine

Data Science

Data Science involves a wide array of technologies, processes, and methodologies to get insights and value out of the data in whatever form, structure, and size it is. A data scientist demands various overlapping skill sets on mathematics and statistics, data mining, machine learning, Computer Programming, Database skills, and Business Domains. Along with that, visualization and presentation skills are also necessary to understand the data and present the information to people. In short, every data science activity hovers around "data," and mining data becomes one of the important skill sets that a data scientist possesses.

Data mining can be defined as the process to understand the business requirement, use statistical analysis and machine learning skills, extract useful information out of the data, and present it to the user. Like software engineering methodologies, there are also several methodologies such as CRSIP-DM and SEMMA (Sample Explore Modify Model Assess) to standardize a data mining project. These processes define the steps to systematically conduct data mining activities. I discuss CRISP-DM methodology, which is mostly followed as an industry-wide standard for data mining projects.

CRISP-DM consists of six phases as illustrated in Figure 1-1. It is a continuous process to improve data mining results after completion of each cycle from business understanding to deployment.

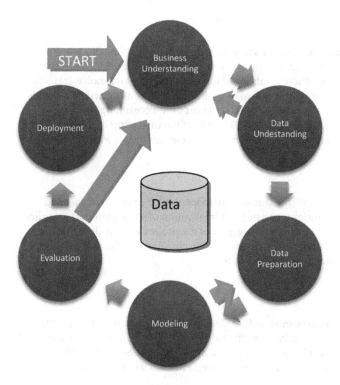

Figure 1-1. CRISP-DM methodology

Business Understanding

Business understanding involves comprehending the business requirements, assessing the goals, and determining the KPIs (key performance indicators), which are going to be affected and improved with the data mining project. If an organization wants to streamline their operations and have a goal to reduce costs associated with their inventory, the team needs to determine the associated KPIs such as cost of days in inventory, inventory turnover ratio, and stock outs ratio tied to the project. This would help them with better solution design and to monitor the success of a project.

Data Understanding

Once the business understanding is established, the data mining team has to validate whether data is sufficiently available and is of good quality to carry out the data mining tasks. The process includes initial data collection, data exploration, and data quality checks. Poor data leads to erroneous findings and is not helpful to business. Sometimes, it might misguide business and lead to poor decisions; so this step is critical, and data needs to be thoroughly examined.

3

Data Preparation

Data preparation is to preprocess the data to a suitable form for building models. Almost any data we encounter in a data science project is not ready to be modeled. Real-world data is often dirty and skewed, there is missing data, and sometimes it is noisy. It is therefore necessary to preprocess the data to make it clean and transformed so it's ready to be run through statistical algorithms. This is the most time-consuming step for any data mining project and consumes almost 60% to 80% of the total effort in a project. However, a simple model on clean data outperforms a complex model with dirty data.

Modeling

Modeling involves the use of statistical models to understand the pattern of data and derive insights out of it. The insights can be just visualization of some patterns or prediction of a future value. It is the most interesting part of the project and involves machine learning as one of its most important components. I discuss machine learning algorithms in the following sections and subsequent chapters.

Evaluation

Once a model is ready, it has to be evaluated to establish its correctness. It is a two-phase process. First, a model is evaluated for its statistical accuracy, that is, whether the statistical assumptions are correct and hold true for other unknown datasets. Second, a model is evaluated to see if it performs as per business requirements and if users truly get some insights out of it.

Deployment

Deployment involves applying and moving the model to business operations for their use. A data mining project is not successful until and unless the model result is used in real time by business. I discuss more about deployment in Chapter 8.

Machine Learning

In a previous section, I discussed machine learning as an important part of the modelling phase. Applying the correct machine learning algorithm is crucial for a data science project's success. We can define machine learning as algorithms that learn and gain experience from the data for identifying patterns to predict the future or to gain some useful information out of it. So if we want to predict the future sales of our organization, we can run the past historical sales data through a machine learning algorithm; and if it has successfully learned the pattern of sales, it will do better at predicting future sales.

Figure 1-2 is an illustration of a machine learning process. The attributes of the input data that goes into a machine learning algorithm is termed as predictors. A machine learning algorithm learns from the underlying data of these predictors and provides output results in a suitable form as per the type of algorithm used. Broadly, machine learning algorithms can be classified into two groups: supervised learning and unsupervised learning.

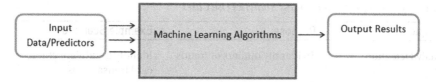

Figure 1-2. Illustration of a machine learning process

Supervised Learning

This group of algorithms works on labeled input datasets, which means they have a specified target or an outcome. The target attribute has some known historical results that can be a continuous numeric attribute, a binary attribute indicating yes/no decisions, or a multiclass attribute with more than two outcomes. Based on these target values, the supervised learning algorithm tries to identify a pattern, establish a relation between the predictors, and then use the derived formula to identify unknown target values from a new dataset. Suppose there is a business requirement to identify the customers who are going to churn in the near future. The input data supplied to the supervised learning algorithms would consider historical data with an indicator to annotate that the customer has or has not churned in the past. Based on this, the machine learning algorithm derives a relationship between predictors for this target value (Churn Yes/No). This formula is then used to predict a customer's churn who is still continuing with the organization. The results provide the organization reasons to take preventive measures to retain customers who are at high risk of leaving. Examples of supervised algorithms include linear regression, logistic regression, Support Vector Machines (SVM), and decision trees. I discuss more about these algorithms in Chapters 5 and 6.

Unsupervised Learning

This group of algorithms works on input data with an unlabeled dataset, that is, data having no target values. The dataset doesn't have any labels and the algorithms identify patterns by deducing structures and the relations of the predictors in the input data. Suppose an organization wants to segment their existing customers for launching a new promotion: they can use unsupervised learning algorithms to identify common patterns across the customer base and accordingly design their promotion activity. Examples of unsupervised algorithms include clustering and association rule mining. I discuss more about these algorithms in Chapters 3 and 4.

Table 1.1 shows some of the business analytics use cases that commonly employ statistical and machine learning algorithms.

Table 1-1. *Commonly Used Business Analytics Use Cases*

Technique	Purpose	Example Scenario
Anomaly Detection	To identify outliers or frauds	Identify fraudulent credit card transactions
Association Rule Analysis	To find frequent patterns in a dataset	Understanding what products go together frequently into a market basket for customers
Attribute Importance	To identify attributes that affect an outcome	Identify the most important factors that influence a purchasing decision
Classification	To predict or categorize new data based on a set of categorized historical dataset; the labeled dataset that goes in to these algorithms has a binary or multiclass target value	Predict the category of a new product that arrives at a retail store
Clustering	Grouping individuals with similar characteristics	Group together customers with similar purchasing patterns
Regression	Predict an outcome based on past historical data; the labeled dataset has a continuous target variable	Predict the sales of a product

Getting Started with Oracle Advanced Analytics

Oracle Advanced Analytics helps organizations to leverage data mining from both structured and unstructured data in the database kernel without any overhead on the client systems or any external integration with other predictive analytics tools. The following are some of the advantages of Oracle's Advanced Analytics platform:

> *Eliminate the data movement*: The data mining algorithms are executed in the database server itself. This eliminates the need to transfer the data to some other server or tools. This also ensures better data quality and minimum latency in data movement.

Deliver security: Data resides in the oracle database and inherits the most trusted security features of the oracle database.

Ease of data preparation: Data preparation, which is the most difficult and time-consuming part of a project, is eased by Oracle. Oracle Advanced Analytics provides automatic data preparation functionality and easy to use APIs (application program interfaces) for custom transformations to reduce the burden for preparing data. This improves the agility of implementing a data science project.

Tight integration to Oracle's technology stack: Being an in-database advanced analytics engine, it is easier to tie the results back to a business operations database quickly and in real time. The results and the models can be invoked by SQL queries, making it quite flexible to use. Oracle Advanced Analytics has already powered analytics capabilities to Oracle's ERP (enterprise resource planning) and its Business Intelligence tools.

Model deployment: Model deployment into a business operations database is simplified and faster, as its data mining programs are coded in SQL or PLSQL, the languages understood by the oracle database. It also supports Predictive Model Markup Language, an open source standard for porting statistical models, which makes it flexible enough to be used in deployment or in any other analytics server.

High data availability and scalability: As the mining algorithm sits at the source of data, correct and legitimate data is available to perform mining activities. In-database and distributed parallel implementation of machine learning algorithms makes Oracle Advanced Analytics platform highly scalable for Big Data.

Flexible interface options: Oracle Advanced Analytics has interfaces to multiple programming languages. It provides easy to use PLSQL, R, and Java APIs for data miners to work with the language they are comfortable with.

Predictive Queries: For users who don't understand mining algorithms, Oracle Advanced Analytics has few APIs that can create models by themselves from the input data and return results. All mining steps are handled by the API itself. This can also be used by data scientists to quickly verify any insights found out of the data.

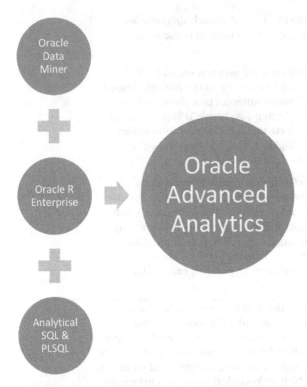

Figure 1-3. *Oracle Advanced Analytics technology stack*

Oracle Advanced Analytics comes up with a powerful combination of analytical SQL and PLSQL packages: Oracle Data Miner and Oracle R Enterprise. Oracle Data Miner is a flagship product from Oracle that provides in-database execution of statistical algorithms to make predictions and discover insights. However, it is packaged with a limited number of algorithms. If we want to execute new advanced algorithms in Oracle Data Miner, it requires coding those algorithms in PLSQL, and it is time consuming. This reduces the flexibility of Oracle Data Miner to implement advanced algorithms quickly. To overcome this challenge, Oracle engineered open source R to Oracle R Enterprise. This engineered feature complemented the limitations of Oracle Data Miner and also made open source R scalable and enterprise ready for Big Data. This also helped Oracle Advanced Analytics to tap into the resources of the huge contributions of the latest libraries by the open source R community. We dive deep into these three components of Oracle Advanced Analytics platform in the following sections.

Oracle Data Miner

In 1999, Oracle purchased Darwin, a data mining product from Thinking Machines. It continued to distribute Darwin while preparing to move these algorithms into the oracle database. In 2002, Oracle Data miner was introduced as Oracle Data Mining 9iR2. Since then, it has gone through many upgrades, changes in its features, and it has introduced new algorithms as well. The current release is Oracle Data Mining 12.1 with many good features I continue discussing in the following sections.

Architecture: Figure 1-4 depicts the architecture for Oracle Data Miner. As seen in Figure 1-4, it has two components:

> ***An enterprise server***: Oracle Data Miner and its repository are installed along with the oracle database. If not, it can be manually turned on in the database.

> ***GUI (graphical user interface) Client***: The Data Miner tab in the SQL Developer is the GUI client for Oracle Data Miner. It is used to visualize the data through various graphs and create data mining workflows. This user interface comes in handy when trying out Oracle Data Miner's features.

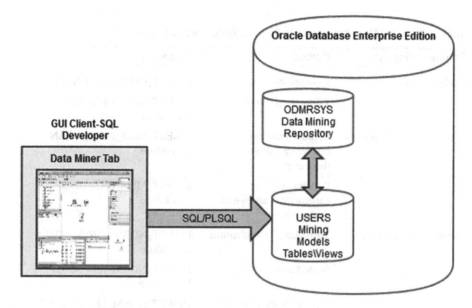

Figure 1-4. *Architecture for Oracle Data Miner*

Oracle Data Miner provides a comprehensive suite of data mining PLSQL packages for implementing data science solutions. Table 1-2 lists, and the following sections brief, three important packages in Oracle Data Miner that provide an array of functions and procedures to drive data science automation and implementations in an organization.

Table 1-2. *Oracle Data Miner Packages*

Data Mining Activity	PLSQL Package
Data preprocessing and transformations	DBMS_DATA_MINING_TRANSFORMATION
Creating, testing, and applying models	DBMS_DATA_MINING
Creating models on the fly	DBMS_PREDICTIVE_ANALYTICS

DBMS = database management system.

DBMS_DATA_MINING_TRANSFORMATION for data preprocessing

DBMS_DATA_MINING_TRANSFORMATION package provides many routines that can be embedded into a data mining project to deal with data problems such as imputing missing values, dealing with outliers, and transforming the data to a suitable form. Using this package, we can automate the data preprocessing step to a maximum extent and help reduce a data scientist's burden to do this task repeatedly for a project.

Table 1-3 lists the routines present in DBMS_DATA_MINING_TRANSFORMATION. I use many of these routines as a part of my solutions for use cases that I will discuss in subsequent chapters.

Table 1-3. *List of Data Preprocessing Routines in Oracle Data Miner*

Preprocessing Type	Method	Procedure
Outlier Treatment	Winsorizing	INSERT_CLIP_WINSOR_TAIL
	Trimming (Clipping)	INSERT_CLIP_TRIM_TAIL Procedure
Missing Value Treatment	Replace missing numerical values with the mean	INSERT_MISS_NUM_MEAN Procedure
	Replace missing categorical values with the mode	INSERT_MISS_CAT_MODE Procedure
Linear Normalization	Min-Max Normalization	INSERT_NORM_LIN_MINMAX Procedure
	Scale Normalization	INSERT_NORM_LIN_SCALE Procedure
	Z-Score Normalization	INSERT_NORM_LIN_ZSCORE Procedure
Binning	Supervised Binning	INSERT_BIN_SUPER Procedure
	Top N Frequency Binning	INSERT_BIN_CAT_FREQ Procedure
	Equi-Width Numerical Binning	INSERT_BIN_NUM_EQWIDTH Procedure
	Quantile Numerical Binning	INSERT_BIN_NUM_QTILE Procedure

Let us take a pause here to understand the purpose for some of these preprocessing methods:

Outlier Treatment: Outliers are values that lie far away from other variables and have some unusual behavior. Presence of outliers may distort the overall distribution of data and give erroneous model results. Based on the number of outliers, they can be either dropped from the analysis or treated accordingly so their presence doesn't affect the overall model results.

> **Winsorizing**: Winsorizing is the process to replace extreme values with less extreme values. It orders the non-null values in numeric columns, computes the tail, values and then replaces the tail values by the specified parameter.

> **Trimming**: Trimming is same as winsorizing except the tail values are just clipped out and set to NULL.

Linear Normalization: Linear Normalization is a process of scaling the variables into a specified range of 0 to 1. It is important to scale the variables when they have different units of measurement to bring them to the same magnitude for comparisons and distance calculations.

> **Min-Max normalization**: In Min-Max normalization, the actual value of a variable (V) is subtracted by its minimum available value (minv) over the difference of its maximum (maxv) and minimum values.

$$V' = \frac{(V - Min_V)}{Max_V - Min_V}$$

> **Scale normalization**: In this type of normalization, the actual value is divided by the maximum of the absolute maximum value and absolute minimum value.

$$V' = \frac{V}{\max(abs(Max_V - Min_V))}$$

> **Z-score normalization**: Here the actual value (V) of a variable is subtracted by its mean value (Meanv) divided by its standard deviation (S.D._v)

$$V' = \frac{V - Mean_V}{S.D._{\cdot V}}$$

Binning: Binning is a process of grouping values into different buckets or bins. Binning forms an important tool in a data science toolbox, as it helps to discretize continuous numeric values or reduce the number of categorical values.

Supervised binning: Supervised binning assigns the bins in a predictor by deriving bin boundaries from decision tree algorithms. It is an intelligent form of binning in which the relationship between the predictor and a target is used to decide the bin values. Values having more predictive power in a predictor are assigned higher bin numbers than values with less or no predictive power. This binning method is suitable to be used in the case of supervised models such as regression or classifications and can be applied on both character and numeric attributes.

Equi-Width Numerical binning: The equi-width bins are formed by dividing the data into k intervals of equal size. The value of "K" is specified by user. This type of binning is only applicable to numerical attributes.

Quantile Numerical binning: Quantile numerical binning is applicable to numerical attributes. Here binning is done by assigning the same number of observations to each bin.

Top N Frequency binning: Top N frequency binning is only applicable to categorical attributes. First, the frequency of each attribute is calculated. Then the top N attributes are assigned individual bin number and the rest of the attributes fall into the (N+1)th bin. Here, N is a user specified number.

Along with these data preprocessing activities, every statistical algorithm requires input data to be in a suitable form to be consumed by it. For example, SVM requires input data to be normalized, whereas Naïve Bayes performs well with discrete data. Oracle has provided automatic data preparation (ADP) functionality in Oracle Data Miner to perform these types of data transformations automatically as required by different algorithms. Table 1-4 lists the data mining techniques that are supported by ADP.

Table 1-4. *Data Mining Techniques That Are Supported by ADP*

Algorithm	Is ADP Supported?
Apriori	No. ADP has no effect on apriori algorithms. Data should be as per apriori algorithm's requirement. We study in depth about Apriori algorithms in Chapter 4.
Minimum Description Length	Yes. Supervised binning is used to bin all attributes.
Generalized Linear Model	Yes. Numerical attributes are normalized.
Decision Tress	No. A decision tree does its data preparation internally.
Naïve Bayes	Yes. Supervised binning is used to bin all attributes.
k-Means	Yes. Outlier sensitive normalization is used to normalize numerical attributes.
Expectation Maximization	Yes. Outlier sensitive normalization is used to normalize numerical attributes.
O-Clusters	Yes. Equi-width binning is used to bin numerical attributes.
Principal Component Analysis	Yes. Outlier sensitive normalization is used to normalize numerical attributes.
Non-Negative Matrix Factorization	Yes. Numerical attributes are normalized.
Singular Value Decomposition	Yes. Outlier sensitive normalization is used to normalize numerical attributes
Support Vector Machine	Yes. Min-max normalization is used to normalize numerical attributes.

O-Clusters = Orthogonal Partitioning Clustering.

DBMS_DATA_MINING for creating, testing, and applying models

DBMS_DATA_MINING package provides an interface to Oracle Data Miner to build a model, test it, and then apply it to a new dataset. It has methods for both supervised and unsupervised mining machine learning algorithms to create models.

The comprehensive suite of data mining algorithms supported by DBMS_DATA_ MINING is as listed in Table 1-5:

Table 1-5. *Data Mining Algorithms in Oracle Data Miner*

Technique	Algorithm	PLSQL Routine in DBMS_DATA_MINING
Anomaly Detection	One class Support Vector Machine	ALGO_SUPPORT_ VECTOR_MACHINES
Association Rule Analysis	Apriori	ALGO_APRIORI_ ASSOCIATION_RULES
Attribute Importance	Minimum Description Length	ALGO_AI_MDL
Classification	Generalized Linear Model	ALGO_GENERALIZED_ LINEAR_MODEL
	Decision Trees	ALGO_DECISION_TREE
	Naïve Bayes	ALGO_NAIVE_BAYES
Clustering	k-Means	ALGO_KMEANS
	Expectation Maximization	ALGO_EXPECTATION_ MAXIMIZATION
	O-Clusters	ALGO_O_CLUSTERS
Feature Extraction	Principal Component Analysis	ALGO_SINGULAR_VALUE_ DECOMP
	Non-Negative Matrix Factorization	ALGO_NONNEGATIVE_ MATRIX_FACTOR
	Singular Value Decomposition	ALGO_SINGULAR_VALUE_ DECOMP
Regression Algorithms	Generalized Linear Model	ALGO_GENERALIZED_ LINEAR_MODEL
	Support Vector Machine	ALGO_SUPPORT_ VECTOR_MACHINES

O-Clusters = Orthogonal Partitioning Clustering.

■ **Note**   I will discuss these algorithms in detail in the following chapters.

DBMS_PREDICTIVE_ANALYTICS to create models on the fly

DBMS_PREDICTIVE_ANALYTICS provides methods to create models on the fly from input data and provide the results. This functionality of Oracle Data Miner is targeted toward users with no statistical modelling or programming skills. We can think it as a form of automated statistics in which a user need not be aware of the mathematical details for building a model. The package internally handles what model to choose when and returns back the output result after necessary computations.

Table 1-6 lists the three methods of the DBMS_PREDICTIVE_ANALYTICS package along with their utility.

Table 1-6. PLSQL Routines in DBMS_PREDICTIVE_ANALYTICS Package

PLSQL Routine	Purpose
PREDICT	Automatically predicts the outcome from the input data. Users need not be aware of the model building steps or prediction algorithms.
EXPLAIN	Returns the important attributes along with their ranks for a specified target value from the input data.
PROFILE	Generate rules for insights from input data.

R Technologies in Oracle

In 2011, Oracle adopted R technologies to form a part of its Advanced Analytics offerings. It provides R integration through four key technologies:

> *Oracle R Distribution*: This is Oracle's distribution for open source R integration. It is available as a free download from the OTN (Oracle Technology Network) web site. This distribution has many performance improvements done by Oracle on open source distribution of R.

> *ROracle*: This is an open source package to interface R with the oracle database. It is publicly available in CRAN (Comprehensive R Archive Network) and is maintained by Oracle.

> *Oracle R connector for Hadoop*: It provides an interface to access HDFS (Hadoop Distributed File System), work with Hive tables, and execute map reduce functions from R. It can be downloaded free from the OTN web site and also forms a part of Oracle Big Data Appliance (BDA).

> *Oracle R Enterprise*: This is an important component of Oracle Advanced Analytics offerings and makes open source R ready for enterprises. In the following section, I discuss at a greater depth its technical components and the advantages that it offers.

Oracle R Enterprise

Oracle R Enterprise is an enterprise version of the highly preferred open source statistical language R. It runs on the database server and integrates R with the oracle database. Traditional R provides a wide variety of machine learning and statistical algorithms, which make it a favorite tool for the data scientist to work on. However, traditional R doesn't scale well with a large dataset. It is a client-based tool that sits at user's machine. This makes data processing slower and not suitable for Big Data. Oracle engineered open source R to Oracle R Enterprise that overcomes the challenges present with open source R. However, using Oracle R Enterprise data, scientists can still continue enjoying the advantages of being a part of large open source R community.

Using Oracle R Enterprise, data scientists can develop scripts/programs that are using R language, execute them, and build models in the database server. The data still remains in the oracle database, and these scripts utilize the ability of database to handle and process millions of records and automate scripts.

Figure 1-5 shows an architectural overview of Oracle R Enterprise. This can be classified into a three-layer component architecture where the

> **1st Layer** is Oracle R distribution or open source R on the user's laptop or desktop.

> **2nd Layer** is the oracle database where SQL generated through the transparency layer is executed. The transparency layer maps R data types to oracle database objects and generates SQL for the executed R expressions.

> **3rd Layer** is spawned R engines of Oracle R Enterprise to execute on the database server under the control of oracle database. Oracle R Enterprise contains statistical libraries that reside at the database server and are executed in the database itself. This embedded execution enables data and task parallelism, XML (extensible markup language), or image extractions and SQL access to R.

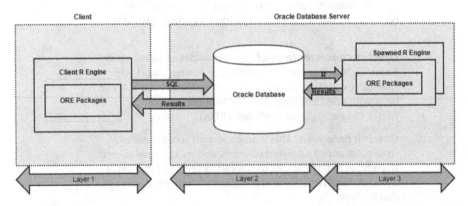

Figure 1-5. *Architecture for Oracle R Enterprise*

ORE also provides a data store to store and retrieve temporary R objects in the database server. If you are working on R on your laptop, you can store the objects in ORE database store. It can be retrieved later to work on. This makes it easy to manage the variety of R objects and makes it independent of your client system.

Advantages of using Oracle R Enterprise

- Ability to execute R code in a database server

- Eliminates the data movement from database to R and vice versa

- Enables SQL access to R

- Can be packaged with PLSQL codes

- Execution can be automated

- Model results can be retrieved as XML files, which can be ported across applications

As we can see from Figure 1-6, Oracle R Enterprise consists of various ORE packages that are installed at both the client and database server. Table 1-7 describes the purpose of these packages. Many packages are reengineered from open source R packages to make them database compatible, while some are Oracle's proprietary packages made to fit it into its advanced analytics offerings.

Table 1-7. *Routines in the ORE Package*

ORE Package	Utility
OREbase	The base package for ORE. Contains methods to convert an object to an ORE object, create constants, database connection functions, and many more
OREstats	Provide methods for statistical tests
OREgraphics	Provide methods for visualization through graphs and plot
OREeda	An exploratory data analysis package that provides various methods for univariate analysis (analyzing a single variable at a time), calculating correlations, summarizing data, ranking values, and many more
OREdm	Exposes proprietary Oracle Data Mining algorithms; this forms a base to integrate the Oracle Data Miner's algorithms to Oracle R Enterprise
OREpredict	Contains functions for generating and scoring model predictions for data present in the oracle database using R models
ORExml	Supports XML translation between R and the oracle database

Apart from these packages, Oracle R Enterprise supports various other open source R packages such as arules, Cairo, DBI, png (Portable Network Graphics), ROracle, statmod, and randomforest.

The ROI (return on investment) of using Oracle R Enterprise lies in its embedded execution of R scripts in the oracle database. It provides two interfaces for this embedded execution: one for users who are comfortable with R and the second one for oracle database users. The variant of interface functions provided in ORE are described following:

Functionality	R Users	PLSQL Users
To invoke stand-alone R scripts	ore.doEval	rqEval
To invoke R scripts with a table as input	ore.tableApply	rqTableEval
For execution of R scripts for row-wise chunked computation	ore.rowApply	rqRowEval
For executing R scripts on data partitions	ore.groupApply	rqGroupEval
To create a R script and store it in the oracle database	ore.scriptCreate	rqScriptCreate
To drop a R script stored in the oracle database	ore.scriptDrop	rqScriptDrop

PLSQL = procedural language/structured query language.

Analytical SQL and PLSQL Functions

Analytics functions in SQL and PLSQL help data scientists and developers to perform various useful tasks using SQL and PLSQL functions that were previously confined to procedural languages. Data wrangling and shaping up the data becomes much easier using these functions. Operations on data is faster, as these functions sit close to the database and can be embedded into the data mining workflows to automate various tasks.

Package DBMS_STATS_FUNC

This PLSQL function packages provides various statistical tasks that can be used to describe and understand the data. The first task data scientists do is to understand the data at hand. For most of the cases, they summarize the data, validate whether the predictors are fitting in a normal distribution, and then carry out various transformations to make it ready to be modelled. Using the functions from DBMS_STATS_FUNC package, which are listed in Table 1-8, a data scientist can automate the mandatory tasks that they need to perform in every data science project. The package provides the summary measures for a dataset and results in a numeric form for distribution tests, which helps in automation.

Table 1-8. *Routines in DBMS_STATS_FUNC*

SUMMARY	Calculates different summary statistics for a dataset such as count of records, minimum and maximum value of an attribute, mean, mode, standard deviation, five quantile values, and top and bottom five values. The summarized values provide data scientists with inputs on the quality of a dataset.
NORMAL_DIST_FIT	Provides tests on how well the attributes in a dataset fit into a normal distribution. SHAPIRO_WILKS test is the default test, although it supports other methods such as KOLMOGOROV_SMIRNOV, ANDERSON_DARLING, or CHI_SQUARED tests. Almost all machine learning algorithms used in business analytics projects are parametric in nature, as parametric statistical tests are more powerful than their equivalent nonparametric tests. Normality is an important assumption for parametric statistical models. It is mostly checked in projects by plotting the data or checking measures such as kurtosis. Using distribution checks provided by this function helps in automation of this activity and takes overhead of manual inspections.
UNIFORM_DIST_FIT	Calculates how well the data fits into a uniform distribution.
WEIBULL_DIST_FIT	Calculates how well the data fits into a Weibull distribution.
EXPONENTIAL_DIST_FIT	Calculates how well the data fits into an exponential distribution.
POISSON_DIST_FIT	Calculates how well the data fits into a Poisson distribution.

Listing 1-1. A Sample Code to Illustrate the Capabilities of This Package

```
DECLARE
summary_values dbms_stat_funcs.SummaryType;
significance number;
BEGIN
dbms_stat_funcs.summary(p_ownername => 'APPS'
,p_tablename => 'TEST_PROD_SALES'
,p_columnname => 'revenue'
,p_sigma_value => 3
,s => summary_values);
dbms_output.put_line('*******');
dbms_output.put_line('Summary statistics of revenue column in TEST_PROD_
                      SALES table:');
dbms_output.put_line('*******');
dbms_output.put_line('Number of records : '||summary_values.count);
dbms_output.put_line('Minimum Value : '||summary_values.min);
dbms_output.put_line('Maximum Value : '||summary_values.max);
```

```
dbms_output.put_line('Variance : '||round(summary_values.variance));
dbms_output.put_line('Stddev : '||round(summary_values.stddev));
dbms_output.put_line('Mean : '||summary_values.mean);
dbms_output.put_line('Mode : '||summary_values.cmode(1));
dbms_output.put_line('Median : '||summary_values.median);
dbms_output.put_line('*******');
dbms_output.put_line('Quantiles');
dbms_output.put_line('*******');
dbms_output.put_line('1st Quantile : '||summary_values.quantile_5);
dbms_output.put_line('2nd Quantile : '||summary_values.quantile_25);
dbms_output.put_line('3rd Quantile : '||summary_values.quantile_75);
dbms_output.put_line('4th Quantile : '||summary_values.quantile_95);
dbms_output.put_line('*******');
dbms_output.put_line('Extreme Count : '||summary_values.extreme_values.count);
dbms_output.put_line('Top Five Values :'||summary_values.top_5_values(1)||',
                                          '||summary_values.top_5_values(2)||',
                                          '||summary_values.top_5_values(3)||',
                                          '||summary_values.top_5_values(4)||',
                                          '||summary_values.top_5_values(5));
dbms_output.put_line('Bottom Five Values : '||summary_values.bottom_5_values(5)||',
                                          '||summary_values.bottom_5_values(4)||',
                                          '||summary_values.bottom_5_values(3)||',
                                          '||summary_values.bottom_5_values(4)||',
                                          '||summary_values.bottom_5_values(5));
dbms_output.put_line('*******');
dbms_output.put_line('Normality Test');
dbms_output.put_line('*******');
dbms_stat_funcs.normal_dist_fit(ownername => 'APPS'
,tablename => 'TEST_PROD_SALES'
,columnname => 'revenue'
,test_type => 'SHAPIRO_WILKS'
,mean => summary_values.mean
,stdev => summary_values.stddev
,sig => significance);
dbms_output.put_line('Significance : '||significance);
END;
```

Here is an output for the Code.

```
PL/SQL procedure successfully completed.
*******
Summary statistics of revenue column in TEST_PROD_SALES table:
*******
Number of records   : 20000
Minimum Value       : 0
Maximum Value       : 1210413.68
Variance            : 2427532790
Stddev              : 49270
Mean                : 33957.336273
Mode                : 0
Median              : 17559.3
*******
Quantiles
*******
1st  Quantile       : 2111.725
2nd Quantile        : 7437.54
3rd Quantile        : 40681.04
4th Quantile        : 118287.18
*******
Extreme Count       : 384
Top Five Values     : 1210413.68,1009957.9,745868.9,711955.2,706259.02
Bottom Five Values  : 0,0,0,0,0
*******
Normality Test
*******
W value : .5914759761737297497910423313969493462734
Significance        : 0
```

Figure 1-6. *Code Output*

Let us understand the results shown in Figure 1-6. The low W value returned from the Shapiro-Wilks test indicates a disagreement with the null hypothesis that the data for the revenue column follows normal distribution. Also, the significance or *p* value is 0, indicating the null hypothesis to be rejected (0.05 being a typical threshold to accept a null hypothesis). We can conclude that the revenue data is non normal.

▦ **Note**   When data is non normal, other distribution tests can be conducted to understand the underlying distribution. Then an appropriate transformation function can be applied to convert it to a normal distribution. Using Oracle's PLSQL APIs and SQL functions, all these tasks can be automated to avoid manual inspections using plots and graphs.

SQL Functions

The analytical functions provided in SQL are grouped into various categories as shown in the following.

Aggregate functions

Average functions are very useful for understanding the data and presenting its summarized view. Knowing the maximum and minimum salary of an employee in a department, average revenue of each product line in an organization, and average customer churn in a year are some of its examples. Some of the widely used aggregate functions such as average, min, max, first_value, and last_value are described in the following section.

Average returns the average value of an attribute. In the following code and the output in Figure 1-7, the average value of revenue per product line is returned:

```
select product_line,avg(revenue)
from TEST_PROD_SALES
group by product_line;
```

⧘ PRODUCT_LINE	⧘ AVG(REVENUE)
1 Outdoor Protection	7693.050708577517314864144912093766648908
2 Golf Equipment	53657.1476810016330974414806750136091453
3 Camping Equipment	40872.074089572886045305204910945875843
4 Mountaineering Equipment	33453.7488519313304721030042918454935622
5 Personal Accessories	30954.7050202569741868271790716518115523

Figure 1-7. Average revenue per product_line

Max returns the maximum value of an attribute. In the following code and the output in Figure 1-8, the maximum value of revenue per product line is returned as a result:

```
select order_method_type, max(revenue)
from TEST_PROD_SALES
group by order_method_type
```

	⇕ ORDER_METHOD_TYPE	⇕ MAX(REVENUE)
1	Special	176832.81
2	Web	1009957.9
3	Telephone	1210413.68
4	Fax	264055.22
5	Sales visit	459086.36
6	Mail	180918.4
7	E-mail	298694.58

Figure 1-8. *Maximum revenue per ORDER_METHOD_TYPE*

Min returns the minimum value of an attribute. In the following code and the output in Figure 1-9, the minimum value of revenue per product line is returned as a result:

```
select order_method_type, min(revenue)
from TEST_PROD_SALES
group by order_method_type
```

	⇕ ORDER_METHOD_TYPE	⇕ MIN(REVENUE)
1	Special	0
2	Web	0
3	Telephone	0
4	Fax	294
5	Sales visit	0
6	Mail	0
7	E-mail	0

Figure 1-9. *Minimum revenue per ORDER_METHOD_TYPE*

CUME_DIST calculates the cumulative distribution of a value in a group of values. In the following code and the output in Figure 1-10, the cumulative distribution of values based on the revenue in each product line is calculated. We can interpret from the results that 60% of the product lines have revenue less than or equal to Golf Equipment.

```
SELECT product_line, sum(revenue), CUME_DIST()
OVER ( ORDER BY sum(revenue)) AS cume_dist
FROM TEST_PROD_SALES
WHERE quarter = 'Q1 2012'
group by product_line;
```

◊ PRODUCT_LINE	◊ SUM(REVENUE)	◊ CUME_DIST
1 Outdoor Protection	6405575.14	0.2
2 Mountaineering Equipment	27392359.63	0.4
3 Golf Equipment	45585426.86	0.6
4 Camping Equipment	101300991.84	0.8
5 Personal Accessories	112544107.06	1

Figure 1-10. Cumulative distribution of total revenue per product_line

First value function returns the first result from an ordered set of values. The following code and Figure 1-11 show the revenue of a product line along with the most unprofitable product line for a quarter:

```
SELECT quarter, product_line,FIRST_VALUE(product_line)
OVER ( ORDER BY tot_revenue ASC ROWS UNBOUNDED PRECEDING) AS unprofitable_
product_line,tot_revenue revenue
FROM (select quarter, product_line,sum(revenue) tot_revenue from TEST_PROD_
SALES where quarter = 'Q1 2012' group by quarter,product_line)
ORDER BY product_line,revenue;
```

◊ QUARTER	◊ PRODUCT_LINE	◊ UNPROFITABLE_PRODUCT_LINE	◊ REVENUE
1 Q1 2012	Camping Equipment	Outdoor Protection	101300991.84
2 Q1 2012	Golf Equipment	Outdoor Protection	45585426.86
3 Q1 2012	Mountaineering Equipment	Outdoor Protection	27392359.63
4 Q1 2012	Outdoor Protection	Outdoor Protection	6405575.14
5 Q1 2012	Personal Accessories	Outdoor Protection	112544107.06

Figure 1-11. Revenue of the most unprofitable product line for a quarter

Last value returns the last result from an ordered set of values. The following code and Figure 1-12 shows revenue of a product line along with the most profitable product line for a quarter:

```
SELECT quarter, product_line,LAST_VALUE(product_line)
OVER (ORDER BY tot_revenue ASC ROWS BETWEEN UNBOUNDED PRECEDING AND
UNBOUNDED FOLLOWING) AS profitable_product_line,tot_revenue revenue
FROM (select quarter, product_line,sum(revenue) tot_revenue from TEST_PROD_
SALES where quarter = 'Q1 2012' group by quarter,product_line)
ORDER BY product_line,revenue;
```

QUARTER	PRODUCT_LINE	PROFITABLE_PRODUCT_LINE	REVENUE
1 Q1 2012	Camping Equipment	Personal Accessories	101300991.84
2 Q1 2012	Golf Equipment	Personal Accessories	45585426.86
3 Q1 2012	Mountaineering Equipment	Personal Accessories	27392359.63
4 Q1 2012	Outdoor Protection	Personal Accessories	6405575.14
5 Q1 2012	Personal Accessories	Personal Accessories	112544107.06

Figure 1-12. *Revenue of most profitable product line for a quarter*

Ranking functions

Ranking functions are used to rank values in a group of values based on the given criteria in the SQL query. These functions are useful for cases where a ranking is to be done based on some parameters: for example, ranking employees within a department; ranking products based on their revenues; and ranking customers based on their monthly Internet usage. Rank and Dense rank are two widely used ranking functions.

Rank provides the rank of a particular value in a group of values. The following code and Figure 1-13 returns the ranks of individual product lines based on their revenue:

```
SELECT product_line,
sum(revenue),
RANK() OVER (order by sum(revenue) DESC) "rank"
FROM TEST_PROD_SALES
group by product_line;
```

Result:

PRODUCT_LINE	SUM(REVENUE)	rank
1 Personal Accessories	267417696.67	1
2 Camping Equipment	236363204.46	2
3 Golf Equipment	98568180.29	3
4 Mountaineering Equipment	62357787.86	4
5 Outdoor Protection	14439856.18	5

Figure 1-13. *Rank of product lines based on their revenue*

Dense rank is same as rank function except that it assigns the same rank to two different records if their values are the same. In the following code and Figure 1-14, the product Hailstorm Titanium Irons for record 180 and 181 are assigned the same rank as their revenue is the same:

```
SELECT product_line,product,
DENSE_RANK() OVER (order by revenue DESC) "rank"
FROM TEST_PROD_SALES
where quarter = 'Q1 2012';
```

Query Result ×

SQL | Fetched 200 rows in 1.342 seconds

	PRODUCT_LINE	REVENUE	rank
172	Personal Accessories	Inferno	172
173	Camping Equipment	Canyon Mule Weekender Backpack	173
174	Camping Equipment	Star Gazer 3	174
175	Golf Equipment	Hailstorm Titanium Woods Set	175
176	Camping Equipment	Star Gazer 2	176
177	Camping Equipment	TrailChef Deluxe Cook Set	177
178	Personal Accessories	TX	178
179	Personal Accessories	Zone	179
180	Golf Equipment	Hailstorm Titanium Irons	180
181	Golf Equipment	Hailstorm Titanium Irons	180
182	Camping Equipment	Star Gazer 2	181
183	Mountaineering Eq...	Husky Rope 200	182
184	Camping Equipment	Star Gazer 2	183
185	Mountaineering Eq...	Granite Extreme	184
186	Golf Equipment	Hailstorm Titanium Woods Set	185
187	Golf Equipment	Blue Steel Max Putter	186
188	Camping Equipment	Star Gazer 3	187

Figure 1-14. *Assignment of the same rank to product Hailstorm Titanium Irons using the DENSE_RANK function*

Statistical functions

Many statistical aggregates, distribution fitting functions, and hypothesis tests are available as functions to be directly applied on data by writing a simple SQL function. This makes it easier for a user to perform the calculations and run hypothesis tests on data sources without the use of any complicated statistical analysis tool. Tests such as the t-Test, F-Test, Chi-square, Kolmogorov-Smirnov Test; statistical aggregates such as variance, standard deviation, correlation, and covariance; and distribution fitting tests such as the Kolmogorov-Smrinov test, Anderson Darling test, Normal, Uniform, Weibull, and Exponential form a part of the statistical SQL functions. The following are some examples of the statistical functions available in oracle database.

Variance is a measure of the spread in values of a variable. It gives a measure of data distribution around the mean, that is, how far the actual value is from the expected value.

Standard deviation is the square root of variance. It standardizes the unit of measure to make the results comparable across variables. The following code and Figure 1-15 are examples of standard deviation and variance:

```
select VARIANCE(revenue), STDDEV(revenue)
from TEST_PROD_SALES
```

⟨ VARIANCE(REVENUE)	⟨ STDDEV(REVENUE)
1 2427532790.43908272360718035901795089754 5	49269.99888815792051364289265751080711 19

Figure 1-15. *Variance and standard deviation of revenue*

Correlation measures the strength of relationship between a pair of variables. It is a standardized measure of covariance, and the value of the correlation coefficient always varies from –1 to +1. There are three widely used types of correlations: Pearson correlation for parametric and linearly related variables, Kendall rank correlation, and spearman correlation for nonparametric measures.

In the following code and Figure 1-16 example, we demonstrate the use of the spearman correlation (corr_s) to calculate the degree of relationship between revenue and gross_margin:

```
select corr_s(revenue,gross_margin) from TEST_PROD_SALES
```

⟨ CORR_S(REVENUE,GROSS_MARGIN)
1   -0.38695659930081111

Figure 1-16. *Correlation of revenue and gross margin*

The *t-test* is used to compare the mean of two groups and determine whether they are statistically different from each other. Depending on the type of test we want to conduct, a t-test can be one-tailed or two-tailed. Two-tailed tests are used to measure the differences between means of two independent samples, and a one-tailed test is used to compare the mean of a population to a target value. The following code and Figure 1-17 are an example of a one-tailed t-test using T_TEST_ONE sql function:

```
SELECT product_line,AVG(revenue) group_mean,
STATS_T_TEST_ONE(revenue, 4000, 'STATISTIC') t_observed
FROM TEST_PROD_SALES
group by product_line;
```

⟨ PRODUCT_LINE	⟨ GROUP_MEAN	⟨ T_OBSERVED
1 Camping Equipment	40872.07408957288604530520491094587584 3	48.3885253182395041744409979690786791905 1
2 Golf Equipment	53657.1476810016330974414806750136091453	39.386570990775701616178335475374850917 56
3 Mountaineering Equipment	33453.7488519313304721030042918454935622	38.831158411849117185040671053770820498 78
4 Outdoor Protection	7693.050708577517314864144912093766648908	16.957248860408859474879866028165917886 04
5 Personal Accessories	30954.7050202569741868271790716518115523	53.078609976949537172676664513777656439 3

Figure 1-17. *Example of a one-tailed t-test*

Bucketing functions

Bucketing functions are useful to bucket or bin the data into distinct groups. For example, we can bucket customers' ages into four distinct groups to analyze their common traits across each age group. WIDTH_BUCKET and NTILE are two widely used bucketing functions.

The *width bucket* function divides the dataset into buckets with an equal interval as specified in the interval size parameter. Anything out of the bucket will be put into an additional bin.

In the following code and Figure 1-18, revenue between $0 and $ 814437 is bucketed into 10 equal bins. Any value falling out of this range will get into the 11th bin.

```
SELECT product, revenue, width_bucket(revenue,0,814437.11,10)
AS bucket FROM TEST_PROD_SALES
where quarter = 'Q1 2012'
and product = 'Star Lite'
```

	PRODUCT	REVENUE	BUCKET
1	Star Lite	1210413.68	11
2	Star Lite	189268.48	3
3	Star Lite	157607.76	2
4	Star Lite	273813.04	4
5	Star Lite	113769.84	2
6	Star Lite	463777.36	6
7	Star Lite	128382.48	2
8	Star Lite	127686.64	2
9	Star Lite	304430	4
10	Star Lite	182310.08	3
11	Star Lite	89415.44	2
12	Star Lite	126642.88	2
13	Star Lite	235193.92	3
14	Star Lite	248414.88	4
15	Star Lite	250850.32	4
16	Star Lite	178135.04	3
17	Star Lite	364272.24	5
18	Star Lite	180570.48	3

Figure 1-18. Width bucket example

NTILE is used to split the data into *n* number of partitions, where *n* specifies the number of buckets to be created. Unlike width_bucket, here the number of buckets remains unchanged, and no additional buckets are created. All the data are fitted into the specified number of buckets. In the following code and Figure 1-19 example, 10 buckets are created based on the revenue of each product for a quarter:

```
SELECT product, sum(revenue), NTILE(10) OVER (ORDER BY sum(revenue) DESC)
AS quartile FROM TEST_PROD_SALES
where quarter = 'Q1 2012'group by product;
```

	PRODUCT	SUM(REVENUE)	QUARTILE
7	Hailstorm Titanium Woods Set	6492651.88	1
8	Canyon Mule Journey Backpack	6087776.43	1
9	Hailstorm Steel Woods Set	5726026.8	1
10	Star Dome	5554895.36	1
11	Star Gazer 3	5477580.87	1
12	Maximus	5438310	1
13	Dante	5381147.05	1
14	Hibernator Extreme	5351965.02	1
15	Venue	5291175	2
16	Canyon Mule Weekender Backpack	5208449.94	2
17	Hailstorm Titanium Irons	4762978.74	2
18	Infinity	4639603	2
19	Lady Hailstorm Titanium Woods Set	4207799.48	2
20	Hibernator	4150123.38	2
21	Trendi	4122789.2	2
22	Hailstorm Steel Irons	4118956.63	2
23	Inferno	4099673.35	2
24	Canyon Mule Extreme Backpack	3815925.35	2
25	Lady Hailstorm Titanium Irons	3812309.91	2
26	Fairway	3675057.75	2
27	Capri	3636585	2
28	TrailChef Deluxe Cook Set	3480411.48	2
29	Hibernator Lite	3458923.96	3
30	Kodiak	3400446.3	3
31	Course Pro Putter	3205371.25	3
32	Lady Hailstorm Steel Woods Set	3057720.48	3
33	TrailChef Single Flame	2910934.32	3
34	Hibernator Self - Inflating Mat	2823148.37	3
35	Canyon Mule Climber Backpack	2763172.04	3

Figure 1-19. NTILE example

Windowing functions

Windowing is a very useful function for any aggregate calculation that involves a range of values or a group of rows. Calculations for moving averages or running totals may require reference to one or more values from preceding or following rows. Here, the window function comes handy. The calculations are carried out across all values from a specified start point to an end point.

In the following code and Figure 1-20 example, a window function is used to calculate the running total of monthly sales of different product lines. The window for the running total calculation is specified as rows between the first row of the results and the current row.

```
SELECT product_line,
SUM(revenue) monthly_sales,
SUM(SUM(revenue)) OVER (ORDER BY product_line
ROWS BETWEEN UNBOUNDED PRECEDING AND CURRENT ROW) running_total
```

```
FROM TEST_PROD_SALES
GROUP BY product_line
ORDER BY product_line;
```

	PRODUCT_LINE	MONTHLY_SALES	RUNNING_TOTAL
1	Camping Equipment	236363204.46	236363204.46
2	Golf Equipment	98568180.29	334931384.75
3	Mountaineering Equipment	62357787.86	397289172.61
4	Outdoor Protection	14439856.18	411729028.79
5	Personal Accessories	267417696.67	679146725.46

Figure 1-20. Example of windowing function to calculate running total

LAG and LEAD functions

The *LAG function* is used to access data from previous rows. It is useful while calculating differences or comparing values between a current record and previous records.

The following code and Figure 1-21 show the current revenue; previous revenue, which is the lag of the current revenue; and the difference in the revenues for the organization. The column revenue_prev is zero for the first column, as its previous value is unknown.

```
SELECT quarter,
sum(revenue) current_revenue,
LAG(sum(revenue), 1, 0) OVER (ORDER BY sum(revenue)) AS revenue_prev,
sum(revenue) - LAG(sum(revenue), 1, 0) OVER (ORDER BY sum(revenue)) AS
revenue_diff
FROM TEST_PROD_SALES
group by quarter
order by quarter desc
```

The *LEAD function* is used to access data from the next rows. It is useful while calculating differences or comparing values between a current record and the following records.

	QUARTER	CURRENT_REVENUE	REVENUE_PREV	REVENUE_DIFF
1	Q3 2012	107737504.97	0	107737504.97
2	Q2 2012	278180759.96	107737504.97	170443254.99
3	Q1 2012	293228460.53	278180759.96	15047700.57

Figure 1-21. Example of LAG function

In the following example, the code and Figure 1-22 show the current_revenue, leading or revenue of the next quarter, and their differential value for product line Outdoor Protection. Notice that the value for next_revenue is zero for Q1 2012, as it is the last column, and the next quarter revenue is unknown.

```
SELECT quarter,
sum(revenue) AS current_revenue,
LEAD(sum(revenue), 1, 0) OVER (ORDER BY sum(revenue)) AS next_revenue,
LEAD(sum(revenue), 1, 0) OVER (ORDER BY sum(revenue))-sum(revenue) AS
revenue_difference
FROM TEST_PROD_SALES
where product_line='Outdoor Protection'
group by quarter
order by quarter desc
```

	QUARTER	CURRENT_REVENUE	NEXT_REVENUE	REVENUE_DIFFERENCE
1	Q3 2012	2132396.32	5901884.72	3769488.4
2	Q2 2012	5901884.72	6405575.14	503690.42
3	Q1 2012	6405575.14	0	-6405575.14

Figure 1-22. *Example of LEAD function*

PIVOT and UNPIVOT functions

The pivot function summarizes or aggregates the data as per a column's value. The pivoted values are shown as column names and the aggregated data as their values. It has the same functionality as pivot in Excel.

In the following code and Figure 1-23, the quantity of sold products for each of the specified product lines is shown. Here the pivot functionality is used to calculate the sum of product quantity for each product_line.

```
SELECT *
FROM (SELECT product_line, quantity
FROM TEST_PROD_SALES)
PIVOT (SUM(quantity) AS sum_quantity FOR (product_line) IN ('Outdoor
Protection' AS Outdoor_Protection, 'Golf Equipment' AS Golf_Equipment,
'Camping Equipment' AS Camping_Equipment
,'Mountaineering Equipment' as Mountaineering_Equipment,'Personal
Accessories' as Personal_Accessories));
```

OUTDOOR_PROTECTION_SUM_QUANTIT	GOLF_EQUIPMENT_SUM_QUANTITY	CAMPING_EQUIPMENT_SUM_QUA...	MOUNTAINEERING_EQUIPMENT_SUM_Q	PERSONAL_ACCESSORIES_SUM_QUANT
2371929	782644	4218093	1558068	5013384

Figure 1-23. *Pivot function example*

Unpivot is used to transform column-based data to individual rows. This is the reverse functionality of the pivot operator.

To demonstrate the unpivot functionality, the test data is created using the following scripts:

```
CREATE TABLE TEST_PROD_QUANTITY (
id NUMBER,
product_line_outdoor NUMBER,
product_code_Golf NUMBER,
product_code_Camping NUMBER
);

INSERT INTO TEST_PROD_QUANTITY VALUES (1, 5404, 3177, 748);
INSERT INTO TEST_PROD_QUANTITY VALUES (2, 114, 304, 321);
INSERT INTO TEST_PROD_QUANTITY VALUES (3, 321, 828, 992);
INSERT INTO TEST_PROD_QUANTITY VALUES (4, 159, 517, 1502);
COMMIT;
```

In the following code and Figure 1-24 example, the product lines are transposed from a columnar format to different rows to indicate different product lines and their respective quantity:

```
SELECT *
FROM    TEST_PROD_QUANTITY
UNPIVOT (quantity FOR product_line IN (product_line_outdoor AS 'Outdoor
Protection', product_code_Golf AS 'Golf Equipment', product_code_Camping AS
'Camping Equipment'));
```

ID	PRODUCT_LINE	QUANTITY
1	Outdoor Protection	5404
1	Golf Equipment	3177
1	Camping Equipment	748
4	Outdoor Protection	159
4	Golf Equipment	517
4	Camping Equipment	1502
2	Outdoor Protection	114
2	Golf Equipment	304
2	Camping Equipment	321
3	Outdoor Protection	321
3	Golf Equipment	828
3	Camping Equipment	992

Figure 1-24. Unpivot function example

Summary

In this chapter, I discussed Oracle advanced analytics' technical stack—Oracle Data Miner and Oracle R Enterprise—and how both these technologies expose hundreds of API to perform data science tasks. In this chapter, I provided examples of various analytical SQL and PLSQL functions that can be useful in a data science project. Apart from these examples, there are other SQL functions that can be used to analyze data. The whole list of analytical functions can be found on the OTN web site. The purpose of this chapter was to introduce you to the world of data science and machine learning and illustrate how the oracle database's analytical queries, Oracle Data Miner and Oracle R Enterprise, can be used to automate the routines in a data science project. The packages and functions form a base for the use cases that we walk through in the next chapters.

CHAPTER 2

■ ■ ■

Installation and Hello World

"The secret of getting ahead is getting started."

—Mark Twain

The first milestone for learning something new is to download the software, install it, and run an example "Hello World" program. It gives us confidence that the platform is all set for us to experiment and motivates us to start our journey to explore more.

In this chapter, I will cover the following topics:

- Installation and configuration of Oracle Data Miner

- Sample Hello World Oracle Data Miner workflow

- Oracle Data Miner components for the SQL developer GUI

- Installation and configuration of Oracle R Enterprise server and client

- Sample Hello World program using Oracle R

- Invoking an ORE script using SQL statements from the SQL developers

---

■ **Note**   It is recommended you install the software and execute the sample programs yourself to have a better understanding of the technology stack. Alternatively, you can download the preinstalled software virtual machine from Oracle Technology Network's Big Data Appliance page.

© Sibanjan Das 2016
S. Das, *Data Science Using Oracle Data Miner and Oracle R Enterprise*,
DOI 10.1007/978-1-4842-2614-8_2

Booting Up Oracle Data Miner

Installation Prerequisites

- Oracle 12c (12.1.0.2) Enterprise Edition/ Oracle 11 Enterprise Edition should be already installed. If not installed, it can be downloaded and installed from Oracle Technology Network Database Download page.

- The latest version of SQL Developer (4.1.3) is installed. If not, it can be downloaded and installed from Oracle Technology Network SQL Developer Download Page.

- Validate whether the Data Miner option is already enabled. This can be done by checking if DATA_MINING option is set to TRUE in the table V$Option. If it is FALSE, ask your DBA (database administrator) to enable the data mining option or enable it running command chopt enable dm from the installed Oracle Database's home bin directory.

- Credentials for SYS user of the Oracle Database.

Optional:

- Sample schemas for Oracle Database are installed. Some of the examples in this book use the tables/views of these schemas.

Installation

A data mining user and Data Miner repository are needed to start using Oracle Data Miner. This can be done using SQL scripts or from SQL developer GUI. A SQL script eases the mundane task of manual configurations each and every time. For those who are not familiar with SQL scripts, SQL developer comes handy. SQL developer is an IDE (integrated development environment) for executing SQL and PLSQL scripts. Oracle Data Miner's GUI and workflows are a part of it and can be easily configured from the SQL developer.

1. Creating a data mining user is the first step in configuring Oracle Data Miner. To start the configuration, open SQL Developer and follow these steps.

2. Connect to user SYS and click on the plus (+) icon located beside the connection as shown in Figure 2-1. This will open up a drop-down list of various options available for user SYS.

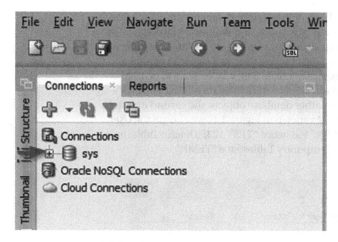

Figure 2-1. SYS user connection tab

Scroll down to the bottom of the drop-down list of options, right-click on the Other Users icon, and select Create User (see Figure 2-2).

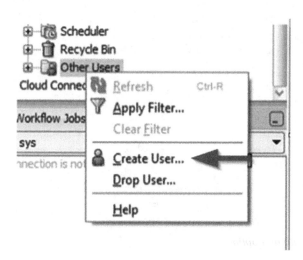

Figure 2-2. Create new user option

3. In the users tab, enter the preferred data mining user name, password, default tablespace, and temporary tablespace. In general, tablespace is a location where all database objects are stored. For a data mining user, you can select either an existing tablespace or create a new tablespace to be used by this user. If you want to store the data mining objects separately from other database objects, the second option is a better choice. As shown in Figure 2-3, we choose the User Name "DMUSER," Password "TEST 123," Default Tablespace "USERS," and Temporary Tablespace "TEMP."

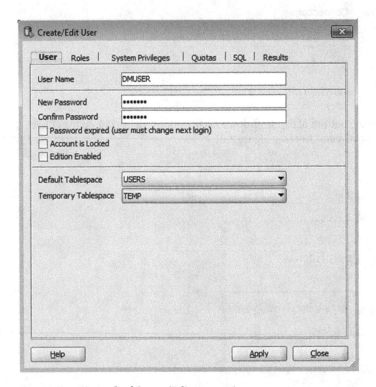

Figure 2-3. User tab of Create/Edit user option

4. Navigate to the System Privileges tab. This tab contains all the privileges to be granted to a user. For experienced Oracle Database administrators, based on their judgement, they can choose the privileges required for that user. However, following (Table 2-1 and Figure 2-4) are some minimum required privileges that are to be granted to a data mining user. To grant the roles, mark the granting option checkbox next to the privileges.

Table 2-1. *List of Privilages to Be Marked*

Privilege

CREATE JOB

CREATE MINING MODEL

CREATE PROCEDURE

CREATE SEQUENCE

CREATE SESSION

CREATE SYNONYM

CREATE TABLE

CREATE TYPE

CREATE VIEW

CREATE ANY DIRECTORY

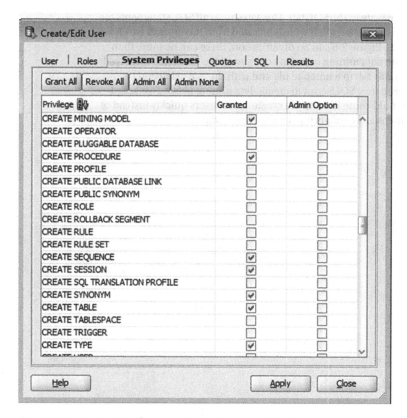

Figure 2-4. *System Privilages marked to be granted to the new user*

5. In the Quotas tab, you can define the amount of space to be allocated to a user in a tablespace. One can limit this space or select unlimited space to be allocated for that user. We have selected unlimited quota for DMUSER (see Figure 2-5).

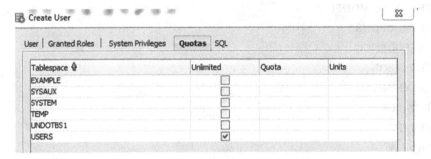

Figure 2-5. *Unlimited Quota marked for tablespace USERS*

6. Navigate to the SQL tab. This tab shows all SQL statements for creating a user that were generated from the options we chose (see Figure 2-6). In an organization, there can be more than one data mining user. One can save the commands displayed in this tab to a notepad file and with the .sql extension to prepare a SQL script to create data mining users. This script would come in handy in creating new users quickly instead of manually configuring the users every time.

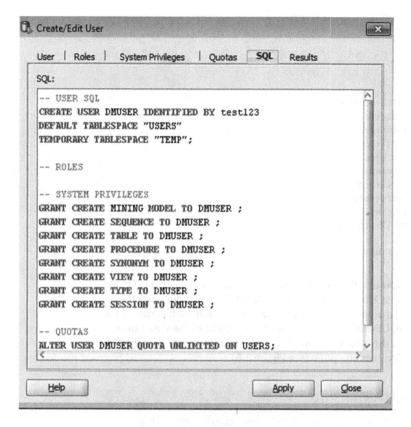

Figure 2-6. *System-generated SQL scripts for creating the user*

7. Continue with the user creation by selecting the Apply button at the lower end of the Create/Edit user screen. The commands get executed, and if everything goes as expected, the success message appears in the screen. In case of any error, the error is displayed.

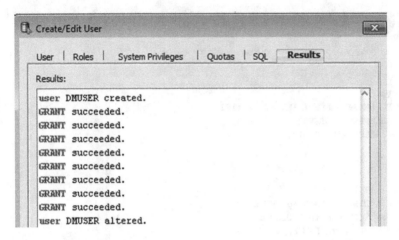

Figure 2-7. Results tab

8. Now that the user is created, it's all set for us to enable Oracle Data Miner and create the data mining repository. Get back to the connections tab, which appears at the left corner of the SQL Developer, or select Connections from the View option on the toolbar. Right-click on Connections and select the New Connection option (see Figure 2-8).

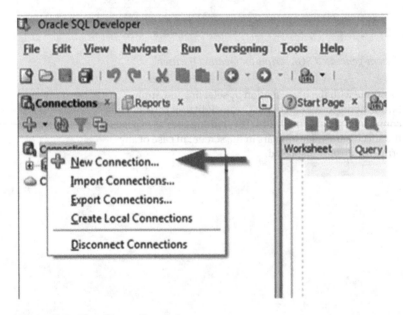

Figure 2-8. New Connection option

9. Enter the details of the user that we recently created. Also, enter the Hostname, Port number and SID (site identifier) of the Oracle Database. You can ask the DBA for these details or you can find it in the TNSNAMES.ORA file. You can provide a connection name to save the connection details in the SQL developer. Select Connect to log in to the database (see Figure 2-9). Alternatively, one can verify the connection details by selecting test option and then connecting to the database.

■ **Note**   TNSNAMES.ORA is a SQL*Net configuration file that has the definitions of database addresses to establish connections with them from a client. It normally resides in the ORACLE_HOME\NETWORK\ADMIN directory.

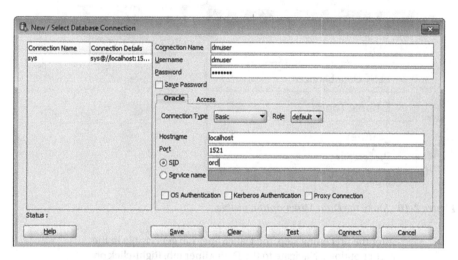

Figure 2-9. *New/Select Database Connection screen*

10. The Data Miner tab is by default invisible. To make it visible, navigate to the Tools section as shown in Figure 2-10 and select Make Visible in the Data Miner option.

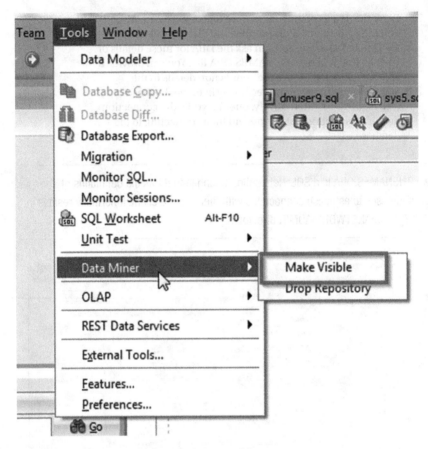

Figure 2-10. Make the Data Miner option visible

11. After the preceding step, the Data Miner tab gets visible in the corner options. Navigate to the Data Miner tab. Right-click on Connections and select Add Connection (see Figure 2-11).

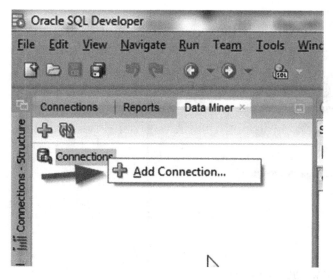

Figure 2-11. *Add Data Miner connections*

12. Select the data mining user, DMUSER, from the Select
     Connection list (Figure 2-12).

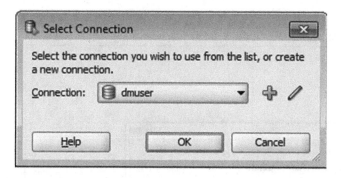

Figure 2-12. *Select Connection screen*

13. Once a user is selected, it appears in the connections tab as
     shown in Figure 2-13.

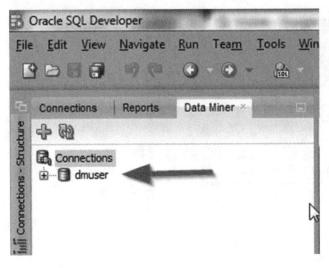

Figure 2-13. *Data Miner DMUSER connection*

14. Select the plus (+) icon beside DMUSER, which will pop up a notification to install the Data Miner repository (Figure 2-14). Click Yes to continue

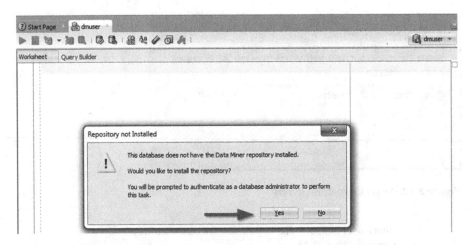

Figure 2-14. *Install the Data Miner repository*

15. As SYS user has SYSDBA credentials, it can be used to install a data mining user. Enter the SYS credentials and continue with the installation (see Figure 2-15).

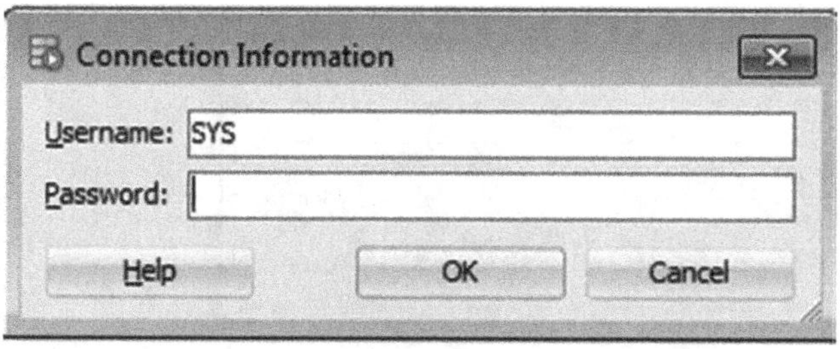

Figure 2-15. Enter SYS user and password screen

16. In the Repository Installation Settings, you have to choose
     the default tablespace and temporary tablespace for the
     ODMRSYS repository account (see Figure 2-16). ODMRSYS
     schema manages the runtime execution of workflows and
     other operations. It also stores the data miner projects and
     workflows that are connected to the database.

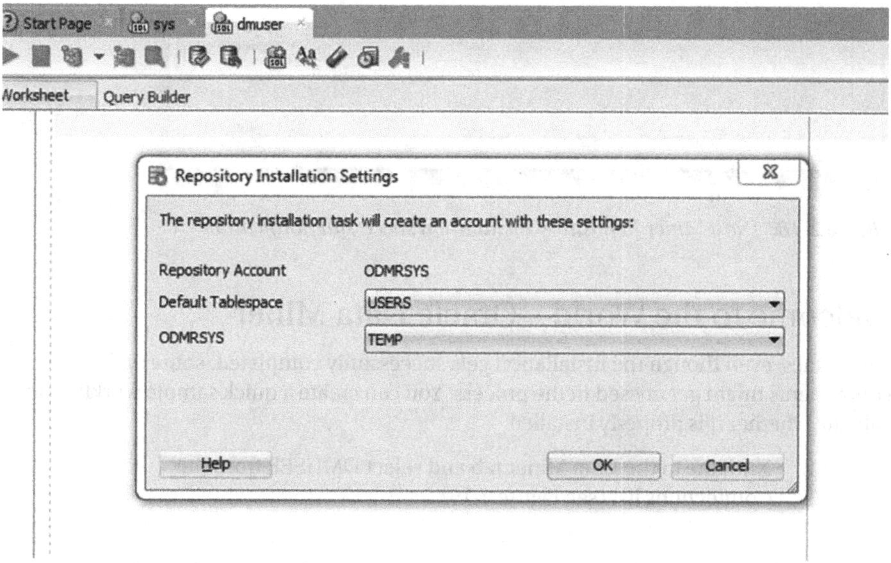

Figure 2-16. Respository Installation Settings

17. In the next step, it shows the dialog box for installing the data
     miner repository (see Figure 2-17). It is the last step of the
     data miner repository creation where the data mining objects
     are created. Click on Start to continue with the repository
     creation.

47

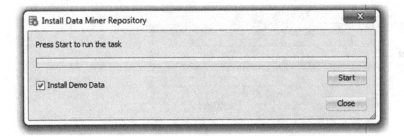

Figure 2-17. *Install Data Miner Repository*

If the Task Completed Successfully message appears as in Figure 2-18, Congratulations!! You have successfully configured Oracle Data Miner. If not, check the log for error details and debug the issue.

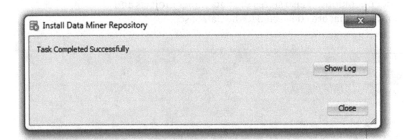

Figure 2-18. *Data Miner Repository installation successful notification*

Welcome to the World—Oracle Data Miner

Sometimes, even though the installation gets successfully completed, some of the components might get missed in the process. You can create a quick sample workflow to validate whether it is properly installed.

1. Navigate to the Data Miner tab and select DMUSER from the Connections list (see Figure 2-19).

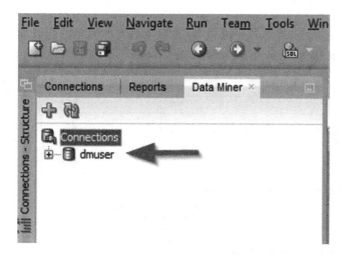

Figure 2-19. *Data Miner dmuser connection*

2. The dmuser connection pops up various options to disconnect the user, add a new user connection, and create new project. It also has options to remove an existing connection or see the user's properties (see Figure 2-20). Select New Project from the list to create a new project.

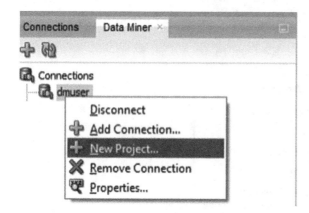

Figure 2-20. *Create New Project option*

3. Enter a project name and project description in the Comment section as shown in Figure 2-21.

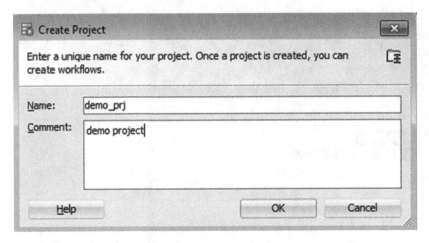

Figure 2-21. Create a project name and comment

4. The project appears in the Connections tab under its schema/
user. Right-click the project and select New Workflow to create
a new data mining workflow (Figure 2-22).

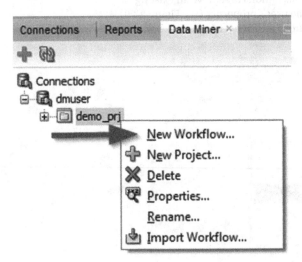

Figure 2-22. New Workflow

5. In the Create Workflow dialog box, enter a name for the
workflow. We have entered it as hello_world as shown in
Figure 2-23.

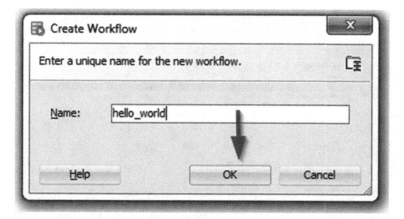

Figure 2-23. *Create New Workflow name*

Once the workflow gets successfully created, the SQL developer IDE opens up various panels specific to data miner. I will discuss those panels in the next section. For the time being, drag the Data Source object from the components panel of Workflow editor to the workflow creation area as shown in Figure 2-24.

■ **Note** The data source node only handles tables or views.

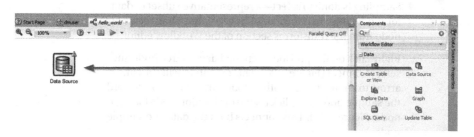

Figure 2-24. *Drag Data Source to the Workflow Editor*

This creates a node to select a data source from the database. Select any of the Tables/Views and click Finish to complete the data source creation as shown in Figure 2-25.

51

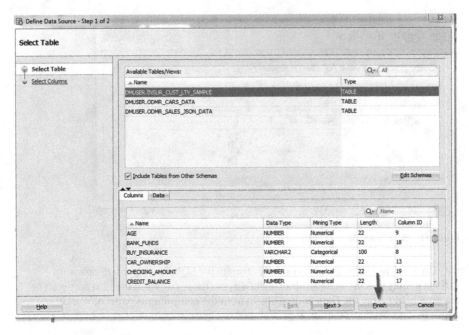

Figure 2-25. Define the data source

6. Next, you would add a sample node for sampling the data. Sampling is done to select—a representative subset of data for modelling when the dataset is very large. Drag the sample node from the transforms section of the Workflow editor.

7. Right-click on the previously created data source node and select Connect (Figure 2-26). This operation will start an arrow to connect it to another node. Drag this arrow toward the sample node and click on the destination node (sample node; Figure 2-27). This connects both the data and sample nodes.

Figure 2-26. *Connect INSUR_CUST_LTV_SAMPLE and Sample node*

INSUR_CUST_LTV_SAMPLE   Sample

Figure 2-27. *Final workflow with connected nodes*

8. To run the workflow, right-click on the Sample node and
    select Run from the menu (Figure 2-28). This operation will
    also execute the prior workflow nodes in the workflow if they
    are not yet executed. Alternatively, you can run it by selecting
    the play button located at top of the workflow worksheet.

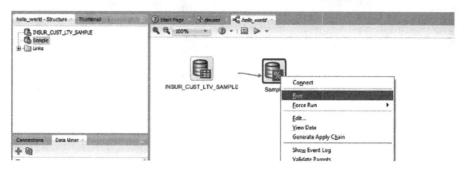

Figure 2-28. *Run the workflow*

53

When this task is successfully completed, it shows a tick in a small green box at each node (Figure 2-29); and if it fails, the node is identified by a cross mark in a red box. This makes it easy to debug the workflow, as it points only to the node that failed, and the event log shows why it failed.

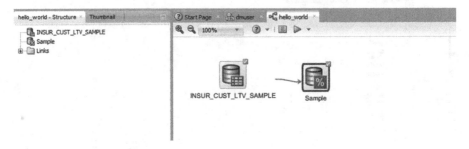

Figure 2-29. Successful execution of the workflow

SQL Developer Components for ODM

The Data Mining option in SQL Developer has various components that help in creating and managing data science projects. I discuss each of those components in this section.

Figure 2-30 shows the Data Miner tab for creating and editing data mining projects and workflows.

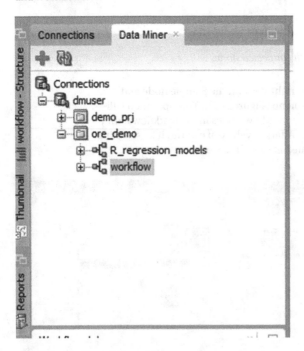

Figure 2-30. Data Miner tab

Figure 2-31 shows the Workflow Jobs dialog box used to monitor the status of all the workflows for a connection. While a job is currently executed, it also shows its progress.

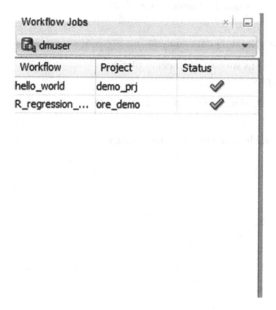

Figure 2-31. *Workflow jobs*

The Components panel (Figure 2-32) has all the functionalities that are supported by Oracle Data Miner to extract the data, transform it, create the model, and apply it to a new dataset.

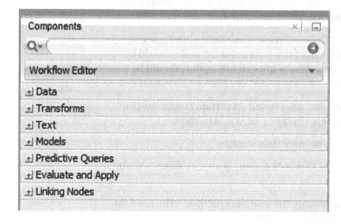

Figure 2-32. *Components panel objects*

Create Table or View: To create a new table/view for storing output results.

Data Source: To define a data source for building model. It can be a table or view.

Explore Data: To conduct exploratory data analysis.

Graph: To create various graphs such as line, scatter plot, bar chart, histogram, and box plot.

SQL Query: To embed SQL, PLSQL, and R scripts into the data mining workflow.

Update Table: To update a table based on a previous node's output.

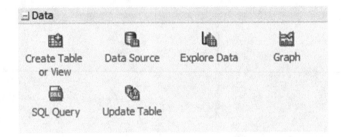

Figure 2-33. *Data section*

Aggregate: Creates aggregated columns (the same as the Group By functionality in SQL statements).

Filter Columns: To select attributes for the data source and for calculating importance of the attributes.

Filter Columns Details: Provides attribute importance details. To run this node, the prior filter columns node should be present, and the attribute importance option should be selected there.

Filter Rows: To filter rows based on certain conditions(the same as the Where condition in SQL statements).

Join: To join two or more data sources.

JSON (JavaScript Object Notation) Query: To extract attributes from a JSON file.

Sample: For specifying the sampling method to be used for creating samples from the data.

Transform: For converting data to a suitable format for modelling. Using this node, you can create new attributes or apply various transformation functions such as log transform to an existing attribute.

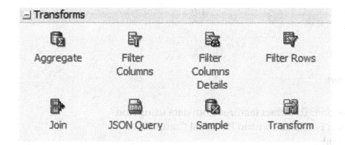

Figure 2-34. *Transforms section*

Apply Text: To apply the text transformations prepared in Build Text node on a new data set.

Build Text: It is used to tokenize the words and calculate their occurrence frequency in a text.

Text Reference: It can be used to reference a text transformation prepared in a current workflow or any other previously saved workflow.

Figure 2-35. *Text section*

Anomaly Detection: To build an anomaly detection model using one class SVM.

Association: To create an association rules model using an Apriori algorithm.

Classification: To create a classification model using decision trees, SVM, Naïve Bayes, and GLM (Generalized linear model).

Clustering: To cluster data using *k*-means, O-Cluster, and Expectation Maximization algorithms.

57

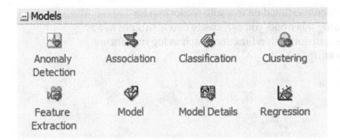

Figure 2-36. *Models section*

> *Feature Extraction*: To extract features from data using Non-Negative Matrix Factorization and Principal Component Analysis methods.
>
> *Model*: To apply an existing model from a data miner repository in the current workflow.
>
> *Model Details*: Shows the various details such as its attributes or rules for a model that were used to build it.
>
> *Regression*: To perform regression analysis on the data using the GLM and SVM algorithms.

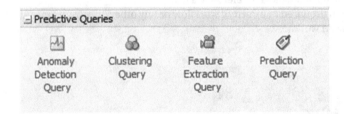

Figure 2-37. *Predictive Queries section*

Predictive queries (see Figure 2-37) are used to create the same models that were discussed in the model section. However, here the user need not be aware of the underlying algorithms. For example, in prediction node for model sections, users are given the option to select algorithms of their choice and parameters to optimize it. In prediction query, the user needs to specify only the data source, and it provides the output. The choice of algorithms and parameters is hidden from the user.

Figure 2-38. *Evaluate and Apply section*

Apply: For applying a model to a new dataset.

Test: To test a classification or regression model using a separate test data source.

Link: To link two different nodes.

Figure 2-39. Linking Nodes section.

ODM Data Dictionary

The data dictionary for Oracle Data Miner provides information on the mining models, the settings, and the attributes of all the created data mining models.

The three database tables in Table 2-2 form part of ODM's data dictionary:

Table 2-2. Mining Tables

*_MINING_MODELS	Provides information about the name of the model, the algorithm used, its size, and the build duration
*_MINING_MODEL_ATTRIBUTES	Provides details of the attributes used in creating the model
*_MINING_MODEL_SETTINGS	Stores the model settings such as the model parameters

■ **Note**   * is to be replaced by ALL/DBA/USER based on user responsibility and access provided.

Boot Up Oracle R Enterprise

Oracle R Enterprise can be downloaded from the Oracle Technology Network (OTN) web site. The latest available version is 1.5. Open source R or Oracle distribution of R can be installed in conjunction with ORE as a client. It is recommended installing Oracle's R distribution, as it is optimized on hardware-specific platforms for faster performance.

Prerequisites

- Oracle 12c (12.1.0.2) Enterprise Edition/ Oracle 11 Enterprise Edition should be already installed. If not installed, it can be downloaded and installed from Oracle Technology Network Database Download page.

- Not all versions of Oracle database, Oracle R Enterprise, and Oracle R Distribution are compatible with each other. It is recommended following the support matrix found on the OTN web site to install the required compatible versions.

- Table 2-3 lists the current Oracle R Enterprise Server Support Matrix and the stack used for this book.

Table 2-3. *Oracle R Enterprise Server Support Matrix*

Oracle R Enterprise	Open Source R or Oracle R Distribution	Oracle Database (see Note)
1.5	3.2.x	11.2.0.4, 12.1.0.1, 12.1.0.2

- Download installation executables for Oracle R Distribution, Oracle R Enterprise server, client, and supporting packages from Oracle Technology Network R Executables Download Page

- Verify the following paths in environment variables

    PATH: Add your R directory such as C:\Program Files\R\R-3.2.0\bin\x64

    R_HOME: Add your R home directory, which looks like C:\Program Files\R\R-2.13.2\bin\x64

    ORACLE_HOME: Add you Oracle home directory, which looks like D:\app\sddas\product\12.1.0\dbhome_1

    ORACLE_SID: Add your oracle database SID such as ORCL

You will first install Oracle R Distribution. It is the Oracle-supported distribution of Open Source R software. This acts as a client for executing R and ORE (Oracle R Enterprise) scripts.

Oracle R Distribution

Run the Oracle R Distribution executable to kick-start its installation. The welcome screen shows the distribution of R that is to be installed. If you have downloaded the right version, click Next to continue with the installation (see Figure 2-40).

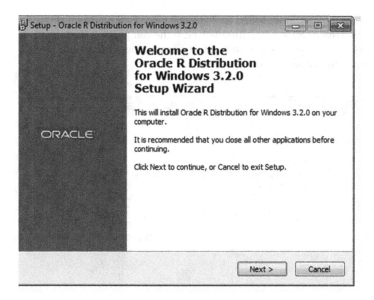

Figure 2-40. *Oracle R Distribution setup*

The next screen shows the license acceptance; accept the license based on the conditions and continue. Choose the destination folder where it is to be installed (see Figure 2-41).

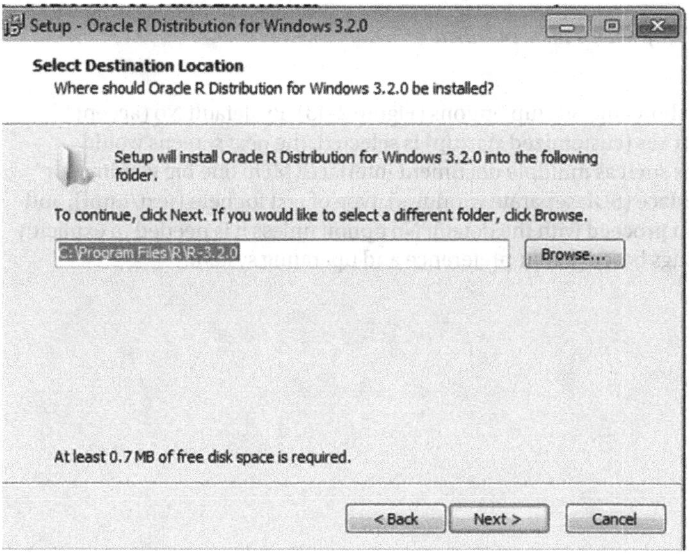

Figure 2-41. *Select installation folder*

The next screen lists the components that are to be installed (Figure 2-42). If you are an advanced user or have open source R installed, clear the components that already exist in your system. However, if you are new to R, select all components for 64-bit User installation and continue by clicking Next.

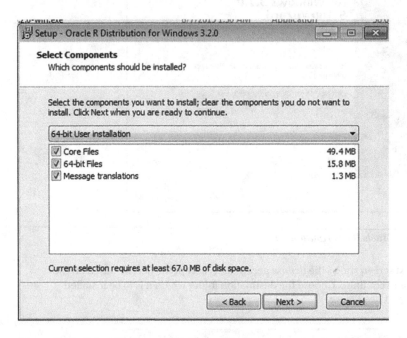

Figure 2-42. *Select Components to be installed*

The next screen shows the Startup options (Figure 2-43). By default No (accept defaults) is selected. If Yes (customized startup) is selected, the next screens would ask for display options such as multiple document interface (MDI; one big window) or single document interface (SDI; separate windows), type of text for help (text/html), and internet proxy. We can proceed with the default No option unless it is needed to explicitly alter installation settings based on our preference and operating system.

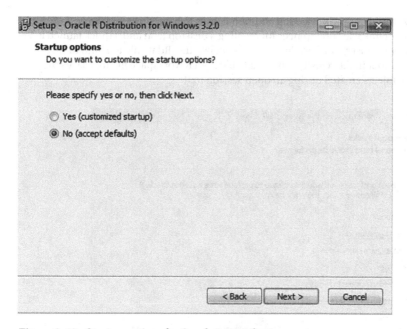

Figure 2-43. *Startup options for Oracle R Distribution*

The next screen shows the setup options for the program's shortcuts (Figure 2-44). The default folder name is R. However, you can change it to some other name of your choice.

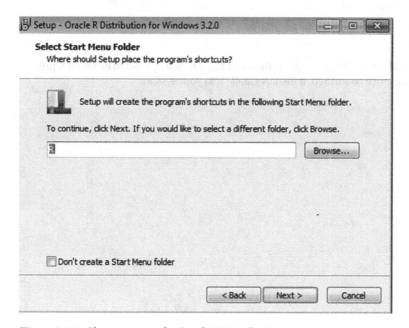

Figure 2-44. *Shortcut name for Oracle R Distribution*

The next setup option is selecting additional tasks to be performed (Figure 2-45). Here, you can define your preferences for creating a desktop icon and a quick launch icon for Oracle R. The registry entries allow you to associate the .Rdata file to R and to save the version number in the registry. The .Rdata file is an R workspace file that saves the function and objects created during an open session in R.

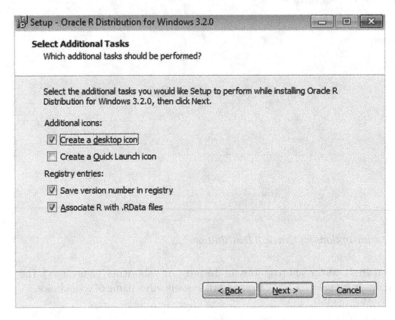

Figure 2-45. *Additional Tasks for installation*

In the final screen (Figure 2-46), it shows all the selected options to be verified prior to installation. Once confirmed, it installs Oracle R in your system.

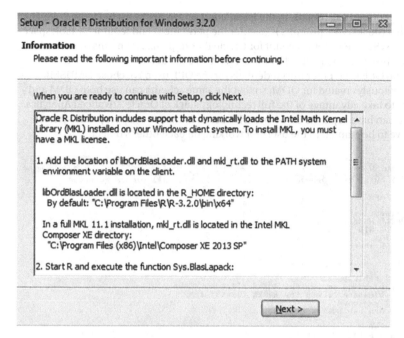

Figure 2-46. Final installation checklist

Next, we need to install Oracle R Enterprise Server. It has libraries that augment Oracle Database to accept requests from Oracle R Enterprise Client and support embedded R script execution in the database.

Oracle R Enterprise Server Installation

Unzip the downloaded zipped ORE packages from the Oracle Technology Network R Executables Download Page. Verify supporting and server folders are in the same directory as shown in Figure 2-47. Install the ORE server by executing the server batch file.

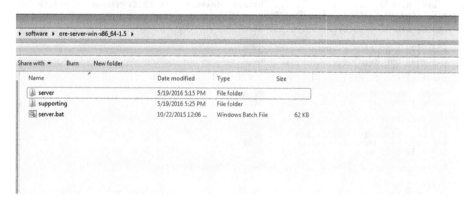

Figure 2-47. Downloaded ORE server and supporting packages

65

Enter the required user details in the installation screen as shown in Figure 2-48. The first option that the installer prompts is to enter the permanent and temporary tablespace for RQSYS. RQSYS is the system account for Oracle R Enterprise. It contains the metadata, functions, and procedure to be used by the server. Once the details are entered, it prompts for the ORE user. Enter the credentials for the ORE user. We chose DMUSER, which was previously created for ODM, so that the same schema can use both ODM and ORE together to take advantage of the full technology stack of Oracle Advanced Analytics. However, you can have separate ODM and ORE users. In that case, appropriate system privileges have to be granted to the users for working on each other's object.

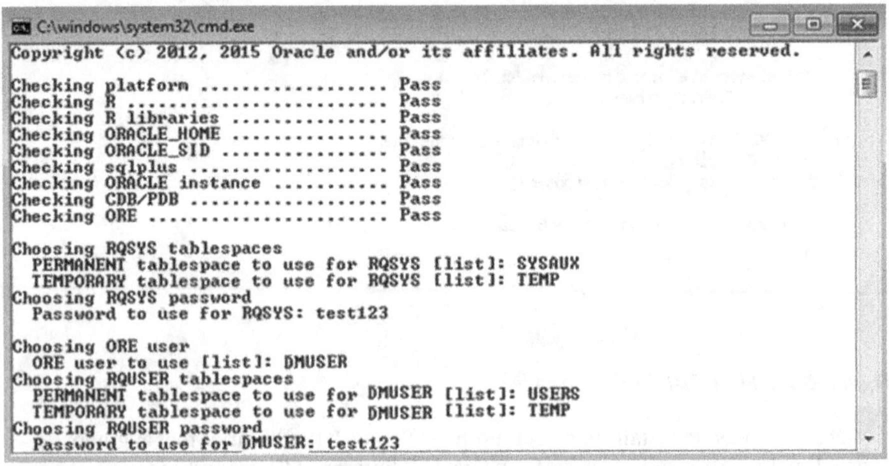

Figure 2-48. *ORE installation screen*

The next screen (Figure 2-49) displays the summary of details that were entered by us. If everything is as per plan, press Enter to continue.

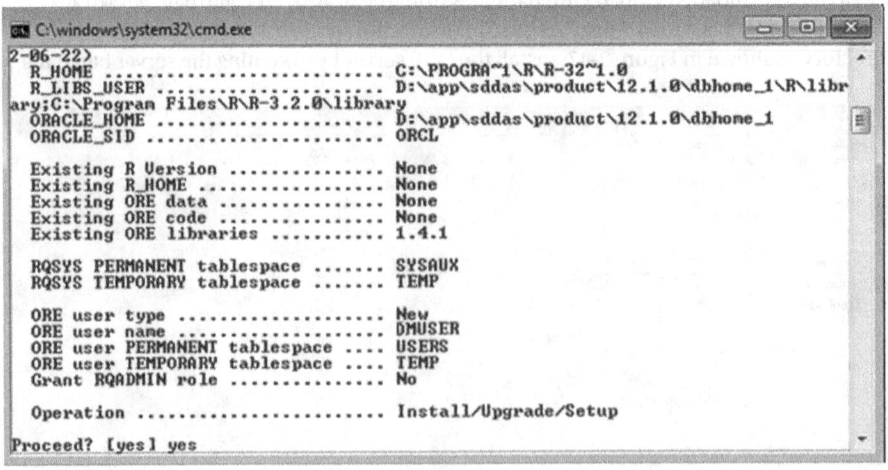

Figure 2-49. *Final ORE installation checklist*

> ■ **Note**   Use server-help in the command prompt to see various options to update the
> user-related information, setup new users, or update ORE.

When the server files are successfully installed, the successful messages "Pass" and
"Done" appear at the bottom of the screen (see Figure 2-50).

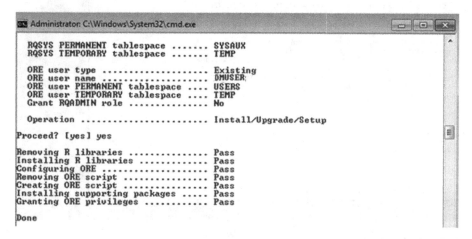

Figure 2-50. ORE server installation successful messages

> ■ **Note**   Verify if the RQADMIN role is granted to DMUSER. If not, grant it by executing
> server RQADMIN from the server.bat location or use SQLPLUS command GRANT RQADMIN
> to DMUSER.

Next, we need to install the client and supporting packages in the system to facilitate
the Oracle R client system in communicating with the ORE Server. We have installed Oracle
Database and ORE server in a single system. However, in organizations, there can be a
single server and multiple clients. In those cases, the following steps have to be carried out
on every client machine that needs to execute ORE scripts and access the ORE server.

Oracle R Enterprise Client Installation

Run the Oracle R distribution executable from the directory that was entered while
installing Oracle R Enterprise. If you are using open Source R as a client interface, then
locate the executable from its installation directory. The executable in both cases is
located in the BIN folder of the installed software directory. R Graphical User Interface
looks like the one Figure 2-51.

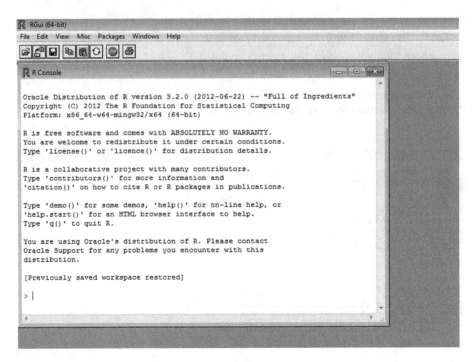

Figure 2-51. R GUI

■ **Note**   Replace the packages path with the corresponding path of your system's file locations.

To install the client files, you can select either of the two following options:

1. Install packages one by one from the packages menu of the toolbar as shown in Figure 2-52.

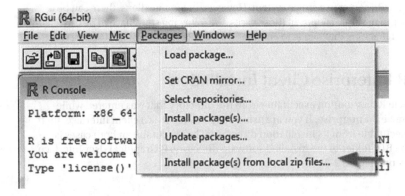

Figure 2-52. Install packages option from R GUI

Select the ORE_1.5 package and choose Open as shown in Figure 2-53.

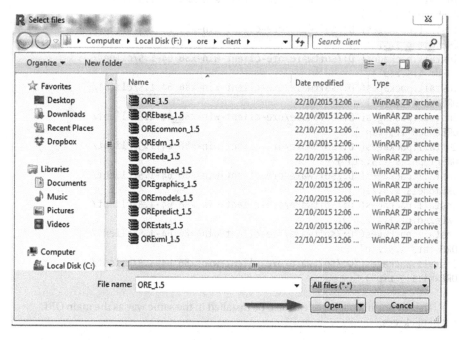

Figure 2-53. *Choose client file for installation*

When the package is successfully installed, package 'ORE' successfully unpacked and MD5 (Message Digest algorithm 5) sums checked messages appear on the screen (see Figure 2-54).

```
[Previously saved workspace restored]

> utils:::menuInstallLocal()
package 'ORE' successfully unpacked and MD5 sums checked
> |
```

Figure 2-54. *Successful package installation message*

This process has to be repeated for all remaining client packages listed in Figure 2-53.

2. Execute the following script for installing all the required client packages together at once. This is the recommended option, as it saves a lot of effort compared to option 1.

```
install.packages("D:/software/ore-client-win-x86_64-1.5/client/ORE_1.5.zip")
install.packages("D:/software/ore-client-win-x86_64-1.5/client/
OREbase_1.5.zip")
install.packages("D:/software/ore-client-win-x86_64-1.5/client/
OREcommon_1.5.zip")
install.packages("D:/software/ore-client-win-x86_64-1.5/client/
OREdm_1.5.zip")
install.packages("D:/software/ore-client-win-x86_64-1.5/client/
OREeda_1.5.zip")
install.packages("D:/software/ore-client-win-x86_64-1.5/client/
OREembed_1.5.zip")
install.packages("D:/software/ore-client-win-x86_64-1.5/client/
OREgraphics_1.5.zip")
install.packages("D:/software/ore-client-win-x86_64-1.5/client/
OREmodels_1.5.zip")
install.packages("D:/software/ore-client-win-x86_64-1.5/client/
OREpredict_1.5.zip")
install.packages("D:/software/ore-client-win-x86_64-1.5/client/
OREstats_1.5.zip")
install.packages("D:/software/ore-client-win-x86_64-1.5/client/
ORExml:1.5.zip")
```

The supporting packages can also be installed in the same way as the main ORE client packages.

Install Supporting Packages for the ORE Client

Execute the following script to install supporting packages in the client machine.

```
install.packages("D:/software/ore-supporting-win-x86_64-1.5/supporting/
arules_1.1-9.zip")
install.packages("D:/software/ore-supporting-win-x86_64-1.5/supporting/
Cairo_1.5-8.zip")
install.packages("D:/software/ore-supporting-win-x86_64-1.5/supporting/
DBI_0.3.1.zip")
install.packages("D:/software/ore-supporting-win-x86_64-1.5/supporting/
png_0.1-7.zip")
install.packages("D:/software/ore-supporting-win-x86_64-1.5/supporting/
randomForest_4.6-10.zip")
install.packages("D:/software/ore-supporting-win-x86_64-1.5/supporting/
ROracle_1.2-1.zip")
install.packages("D:/software/ore-supporting-win-x86_64-1.5/supporting/
statmod_1.4.21.zip")
```

When the files are successfully installed, the package is successfully unpacked, and MD5 sums checked messages appear on the screen as in Figure 2-55.

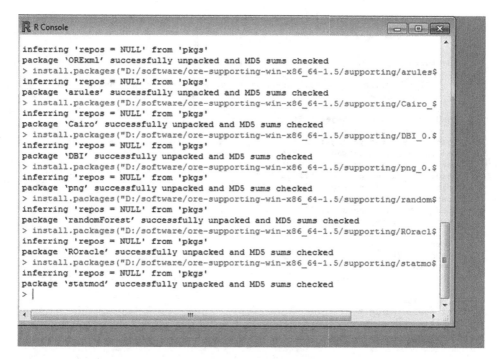

Figure 2-55. Supporting files for ORE client installation

Welcome to the World of Oracle R Enterprise

Execute your first ORE code and verify the installation using the following snippets (see Figure 2-56).

```
#1. Load the  ORE library
library(ORE)
```

```
#2. Connect to the database using ore.connect(). Replace the parameters with
your Database #credentials
ore.connect(user="DMUSER",sid="ORCL",host="localhost",
password="sibanjan123","1521", all=TRUE)
```

```
#3. Invoke standalone ORE function using ore.doEval()
test <- ore.doEval(
function() {
"Hello World!"
}, ore.connect = TRUE);
```

```
#4. Print the result
print(test)
```

```
R R Console                                                     [ _ ][ □ ][ x ]

 The following objects are masked from 'package:base':

     cbind, data.frame, eval, interaction, order, paste, pmax, pmin,
     rbind, table

 Loading required package: OREembed
 Loading required package: OREstats
 Loading required package: MASS
 Loading required package: OREgraphics
 Loading required package: OREeda
 Loading required package: OREmodels
 Loading required package: OREdm
 Loading required package: lattice
 Loading required package: OREpredict
 Loading required package: ORExml
 > ore.connect(user="DMUSER",sid="ORCL",host="localhost",password="sibanjan123",$
 > test <- ore.doEval(
 + function() {
 + "Hello World!"
 + }, ore.connect = TRUE);
 >
 > print(test)
 [1] "Hello World!"
 > |
```

Figure 2-56. *The "Hello World!" program using ORE script*

Execute the following code snippets to check embedded database execution of the R scripts from PLSQL (also see Figure 2-57).

```
//1. Create a R script to be stored in the database's R repository
identified by script name test_plsql

begin
sys.rqScriptCreate
('test_plsql','function() {"Hello World!"}');
end;
/

//2. Invoke the previosuly stored R script(test_plsql) using rqEval and
return the result as an XML file

select name, value
from table(
1rqEval
(
NULL,
```

```
'XML',
'test_plsql'));
```

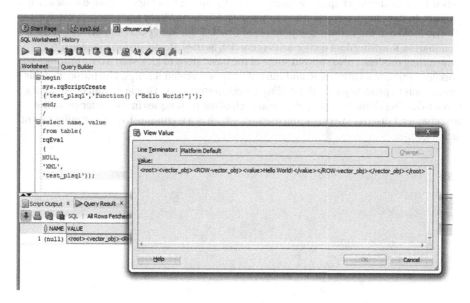

Figure 2-57. *Invoke ORE script using SQL statement*

ORE Data Dictionary

The data dictionary for ORE (Table 2-4) provides information on the R scripts stored in the database. Also, it has database tables that can be queried for managing the datastores.

Table 2-4. *ORE Data Dictionary*

rq$script	Table that lists all the R scripts stored in the database. This table has to be queried from the SYS schema
sys.rq_scripts	A view that lists the R scripts that are present for RQSYS schema
RQUSER_DATASTORELIST	A view that lists all datastores for the current schema
RQUSER_DATASTORECONTENTS	A view that provides object-level information for all datastores in the current schema

73

Summary

This was a foundation chapter for using Oracle Advanced Analytics for your data science projects. In this chapter, you learned to install Oracle Data Miner and Oracle R Enterprise. We also looked at the various components of Oracle Data Miner GUI and discussed data dictionaries. Knowing data dictionaries is crucial, as they are needed to manage the data mining models. I also showed some of the basic steps to kick-start a data mining workflow, execute a ORE script, and run a PLSQL embedded R script. The purpose of this chapter was to provide you with building blocks to start using Oracle Advanced Analytics. In the following chapters, I deep dive into each of the techniques in data science. In the next chapter, I discuss the clustering method, which is a technique to group similar sets of data together.

CHAPTER 3

■ ■ ■

Clustering Methods

Mobile and tablet devices and easy access to the Internet are making the world go digital. Retail business is no longer confined to brick and mortar stores. There are multiple channels through which business is carried out. Some consumers prefer to shop using mobile or web applications; but they never stop visiting brand stores. At each touch point, customers leave their footprints and are captured through different attributes. There are thousands of attributes for millions of customers. It is tough to understand each of them personally. Tracking these customers and segmenting them into different groups is becoming increasingly difficult. So how do we segment our customers to target them for a promotional activity? There is an opportunity for you to use clustering techniques to segment your customers. Clustering techniques can group attributes into a few homogeneous segments where data within each group are similar to each other and different across groups. It is an unsupervised learning process finding logical relationships and patterns from the structure of the data. Clustering finds its application in data mining, information retrieval, image segmentation, and grouping similar web pages and is used across industries. It can be utilized for

- Customer segmentation where a set of homogeneous customers are grouped together

- Industry analysis to find a group of similar firms

- Social network analysis to find a coherent group of friends

- Grouping text documents to assign them a category

- Detect fraud in transactions

And much more. . . .
In this chapter, I will discuss the following topics:

- Concepts and approaches for cluster analysis

- *k*-means algorithm fundamentals

- *k*-means algorithms in Oracle Advanced Analytics

- Clustering rules evaluation metrics

- Creating clusters using Oracle SQL and PLSQL APIs

© Sibanjan Das 2016
S. Das, *Data Science Using Oracle Data Miner and Oracle R Enterprise*,
DOI 10.1007/978-1-4842-2614-8_3

- Creating clusters using Oracle R Enterprise
- Creating clusters using Oracle SQL Developer GUI
- Walk through a clustering case study—Customer segmentation

Clustering Approaches

There are various methods to cluster data together. It can be as simple as filtering data based on certain attributes such as grouping customers based on their age group or using mathematical algorithms to allow data to form clusters automatically. Sometimes both the approaches can be employed together to provide a more robust solution. Later in this chapter, my case study is based on a scenario where both these clustering approaches are used simultaneously to formulate a customer segmentation strategy for an organization.

The following is a brief of the types of clustering algorithms and their implementation in Oracle Advanced Analytics:

1. ***Hierarchical method***: In this method, a hierarchy of clusters is formed based on the distance between pairs of data points. It starts with one record and iteratively pairs them together until a single cluster is found. The clusters form a tree-like structure and can be visualized through a dendrogram plot. The process of finding clusters in this method is resource intensive but allows one to choose the number of clusters by visualizing the cluster results.

   Oracle Advanced Analytics doesn't have this method as a separate algorithm. However, it has implemented it as an enhancement for all of its clustering algorithms. For example, users can take advantage of the clusters formation process of a distance-based clustering algorithm such as k-means and visualize the clusters in a tree structure as in a hierarchical method.

2. ***Density-based technique***: This method finds the cluster by using specific probability distribution of the data points. The idea here is to continue growing the clusters as long as the density in the neighborhood exceeds a specified threshold. The high-density regions are marked as clusters separated from low-density regions, which might be noise. Noise is a random error or variation in a data set that is statistically uncertain and cannot be explained.

   Oracle Advanced Analytics implements the expectation maximization algorithm, a variant of the density technique wherein the expectation step computes the likelihood of cluster membership based on the overall probability of data. In the maximization step, it tries to maximize the overall probability or likelihood of the clusters.

3. ***Grid-based method***: This method operates by dividing the attributes of a data set into hyper-rectangular grid cells. It then eliminates low-density cells that are below a specified threshold parameter. The adjacent high-density cells are then combined until the objective function is minimized. The resulting cells are interpreted as clusters.

   Oracle implements its proprietary grid-based algorithm, O-Cluster, which creates orthogonal partitions for the attributes in the data set to form grids. The algorithm does not use all the data passed as its input. It reads the data in batches, and based on statistical tests decides if there are more clusters to be discovered. This causes its processing time to be faster than any other algorithm.

4. ***Partition-based methods***: In a partitioning method, data is divided into partitions based on the distances between the data points; the k-means algorithm is a widely used partitioning type of clustering. In a partition-based clustering method, choice of the appropriate distance function influences the shape of clusters. Euclidean, cosine distance, and fast cosine distances are three distance functions available in Oracle Advanced Analytics for k-means algorithms. Euclidean distance is mostly sensitive to the scale of the input vectors. In such cases, one has to normalize the scale of input vectors or choose a scale-insensitive distance measure such as cosine distance.

   Oracle has implemented an enhanced version of the k-means algorithm; it is a popular clustering algorithm and is widely used across industries. For this reason, I will discuss the k-means algorithm in the following section at greater depth. Although I limit this discussion to k-means, the other two algorithms present in Oracle Advanced Analytics can be run in a similar way with some changes in their input setting parameters. The objective here is to introduce you to the methodology used to implement a clustering algorithm using the Oracle Advanced Analytics platform.

Oracle data miner of the Oracle Advanced Analytics platform implements three algorithms—k-means, O-Cluster, and Expectation Maximization. However, they cover all four widely used types of clustering in industries. The hierarchical method is common and implemented for the three algorithms supported in Oracle Advanced Analytics.

The *k*-means Algorithm

The k-means is a simple algorithm that divides the data set into k partitions for n objects where $k \leq n$. In this method, the data set is partitioned into homogeneous groups with similar characteristics. The similarity or dissimilarity is defined by calculating the distance between the centroid and data points. The clusters are formed when the optimization function for the algorithm achieves its objective—less intracluster distances and more intercluster distances.

■ **Note**   Optimization function is used to achieve an algorithm's objective along with minimizing the cost or error associated with learning the training data. For algorithms that are based on some reward system, the optimization function is maximized.

The following steps are involved in traditional k-means clustering:

1. Specify the number of clusters and metrics to calculate the distance

2. Randomly pick the initial centroid per number of clusters

3. For each data point

    a. Calculate the distance between the centroid and data points

    b. Assign each data point to the nearest centroid of a cluster

4. Calculate the new centroid for each cluster

5. Repeat steps 3 and 4 until the centroids in clusters change. Terminate when they remain unchanged in successive cluster reassignments.

Consider the following example in Figure 3-1 where we want to cluster five students into two groups based on their history and geography scores.

History	Geography
84	49
69	33
67	37
78	45
81	49

Figure 3-1. *History and Geography scores*

As this is a two-dimensional data set, we plot the history and geography scores for each student in the x and y axis, respectively. We can visualize two prominent clusters from the graph (see Figure 3-2). Let's run this data through the k-means algorithm and find out the cluster results.

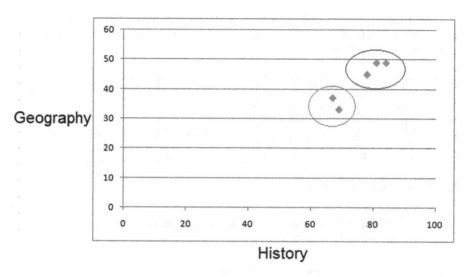

Figure 3-2. *Two-dimensional plot student's scores*

> **STEP 1:** Specify the number of clusters and metrics to
> calculate the distance.

As we intend to find two clusters in our sample data set, we set number of clusters
(k) = 2. Assuming the marks are in the same unit and scale, we choose Euclidean distance
as the metric for calculating distance. Euclidean distance between a point (x1,y1) and the
centroid (cx1,cy2) is given by the following:

$$\text{dist}((x1,y1),(cx2,cy2)) = \text{sqrt}((x1\text{-}cx2)2+(y1\text{-}cy2)2$$

> **STEP 2:** Algorithm finds the initial centroid.

As we are doing the k-means calculation manually, we randomly choose the first two
records to be the initial centroids (see Figure 3-3).

Cluster	Initial Centroid	
	History	Geography
c1	84	49
c2	69	33

Figure 3-3. *Initial centroids*

> **STEP 3:** Calculate the distance between the centroid and each
> data point.

Using the Euclidean formula given in the equation, we calculate the distance of each data point from each cluster. The closest cluster centroid was assigned a cluster for the respective data point (Figure 3-4).

History	Geography	Distance from cluster 1	Distance from Cluster 2	Cluster Assignment
84	49	0	21.9317122	1
69	33	21.9317122	0	2
67	37	20.80865205	4.472135955	2
78	45	7.211102551	15	1
81	49	3	20	1

Figure 3-4. *Cluster assignment*

STEP 4: The new centroid/mean for each cluster is recalculated (Figure 3-5).

Cluster	New centroid	
	History	Geography
c1	81	47.66666667
c2	68	35

Figure 3-5. *New centroid*

STEP 5: The distance between data points and cluster centroids are again calculated. As the cluster assignment doesn't change, we get our two final clusters. If we compare the clusters in the plot and in Figure 3-6, we find the *k*-means algorithm has discovered the same clusters.

History	Geography	Distance from cluster 1	Distance from Cluster 2	Cluster Assignment
84	49	3.282952601	21.26029163	1
69	33	18.9502272	2.236067977	2
67	37	17.60050504	2.236067977	2
78	45	4.01386486	14.14213562	1
81	49	1.333333333	19.10497317	1

Figure 3-6. *Cluster assignment after recalculating new centroids*

k-means in Oracle Advanced Analytics

Oracle has enhanced the functionality of the *k*-means algorithm to drive automation. Some of its features that beat out the traditional *k*-means algorithm are outlined following:

- Builds k-means in an hierarchal fashion

- The traditional K-means algorithm works only on numeric data; in Oracle Advanced Analytics, it is designed to handle both numeric and categorical variables

- Uses probabilistic scoring of data for cluster assignments

- Provides rules that describe data in each cluster

- Provides real-time scoring for cluster membership of a new data set

- As k-means is outlier sensitive, the automatic data preparation (ADP) module for k-means performs outlier-sensitive normalization; to use this feature, ADP should be set to TRUE while configuring the algorithm.

- Using this algorithm, users need not create multiple k-means models to arrive at an optimal value for k; instead, they can decide the value after the first run itself

Clustering Rules Evaluation Metrics

Each cluster is described by certain rules that explain its data. To evaluate the accuracy of the rules, Oracle Advanced Analytics is equipped with two parameters—Support and Confidence:

- Support describes the percentage of records for which the rule holds true.

- Confidence is the probability that a particular record will be assigned to the top rule that describes the cluster.

Parameters to Tune k-means Clustering

As a data scientist, you are required to work on different data sets and might be using the same algorithm for the assignments. However, as per the task and data set, you may need to tweak certain parameters to achieve an optimal number of clusters. Oracle has provided eight parameters to tune the k-means algorithm. It has assigned all parameters with some standard values and can be overridden by specifying the accepted values for each parameter (see Table 3-1).

Table 3-1. *Parameters to Tune k-Means Clustering in Oracle Advanced Analytics*

Parameter	Default Value	Function It Performs
Number of clusters	10	To set the required number of clusters.
Growth factor	2	To allocate memory for holding cluster data; it can range from 1 to 5.
Convergence tolerance	0.01	For setting the error tolerance level; it ranges from 0.001 to 1. The lower the value, the more optimal clusters are formed but with an overhead of higher processing time.
Distance function	Euclidean	To choose any of three distance functions for cluster calculations— Euclidean, cosine, and fast cosine.
Number of iterations	3	To set the maximum number of iterations for the clustering algorithm; the model builds faster with a small number of iterations. However, the result might not be optimal.
Minimum percent attribute rule support	0.1	Used to specify a threshold for attributes to qualify as a rule in the cluster; it ranges from 0 to 1.
Number of histogram bins	10	This algorithm uses the equi-width bin to discretize continuous variables.
Split criterion	Variance	Criteria to split a cluster further; it is used to initialize a k-means cluster. The parameter accepts only one of the two values—variance and size. Split by variance partitions the cluster with most spread out data; split by size partitions the cluster having largest area.

Creating a Cluster Model in Oracle Advanced Analytics

The beauty of using Oracle Advanced Analytics is that its algorithms can be utilized by a business analyst who is not familiar with programming, a database programmer who knows only SQL/PLSQL, and a statistician who is more acquainted with only R language. In this section, I walk you through the different available methods to create a cluster model using PLSQL API, R scripts, and SQL developer user interface (UI).

■ **Note**   The data for the following examples can be found in the book's companion folder. The data set used is the Iris flower data set, which is a multivariate data set introduced by Ronald Fisher. The data set has 50 samples from each of three species of Iris. Four features were measured from each sample: the length and the width of the sepals and petals, in centimeters.

Clustering using SQL and PLSQL

STEP 1: Create a setting table

The setting table is necessary to override the default values for the model-setting parameters. You can skip this step if you are creating the model with Oracle-provided default values.

```
-- Create Settings tables to be used in model building
CREATE TABLE odm_km_settings (
    setting_name  VARCHAR2(30),
    setting_value VARCHAR2(4000));

-- Insert settings
SET SERVER OUTPUT ON;
BEGIN
    INSERT INTO odm_km_settings (setting_name, setting_value) VALUES
    (dbms_data_mining.kmns_distance,dbms_data_mining.kmns_euclidean);

    INSERT INTO odm_km_settings (setting_name, setting_value) VALUES
    (dbms_data_mining.prep_auto,dbms_data_mining.prep_auto_off);

    INSERT INTO odm_km_settings (setting_name, setting_value) VALUES
    (dbms_data_mining.CLUS_NUM_CLUSTERS,3);
    -- Other examples of overrides are:
    -- (dbms_data_mining.kmns_iterations,3);
    -- (dbms_data_mining.kmns_block_growth,2);
    -- (dbms_data_mining.kmns_conv_tolerance,0.01);
    -- (dbms_data_mining.kmns_split_criterion,dbms_data_mining.kmns_variance);
    -- (dbms_data_mining.kmns_min_pct_attr_support,0.1);
    -- (dbms_data_mining.kmns_num_bins,10);
END;
/
```

STEP 2: Create the mining model

To create the clustering model, you need to call the CREATE_MODEL procedure of the DBMS_DATA_MINING package with the parameters shown in Table 3-2.

***Table 3-2.** Parameters for Creating a Clustering Model Using DBMS_DATA_MINING. CREATE_MODEL*

Parameter	Mandatory	Description
model_name	Yes	To assign a meaningful name for the clustering model.
mining_function	Yes	The data mining function that is to be used; for clustering, you need to call the dbms_data_mining. clustering function.
data_table_name	Yes	Name of the data table/view to be used to create the model.
case_id_column_name	No	The unique record identifier for the data set; the primary key of a table can be used as a case id column. If there is no unique ID column, it can be set to null. However, the processing time for building the model might be more.
settings_table_name	No	Name of the table that contains the model settings values. If null is passed to this parameter, the model will process with default values

The syntax for creating new clustering model is as follows:

```
BEGIN
  DBMS_DATA_MINING.CREATE_MODEL(
    model_name          => 'KM_ODM_MODEL',
    mining_function     => dbms_data_mining.clustering,
    data_table_name     => 'XXCUST_IRIS_DATA_V',
    case_id_column_name => 'ID',
    settings_table_name => 'odm_km_settings');
END;
/
```

STEP 3: Check if the model is created successfully (see Figure 3-7)

```
select * from user_mining_models where model_name = 'KM_ODM_MODEL'
```

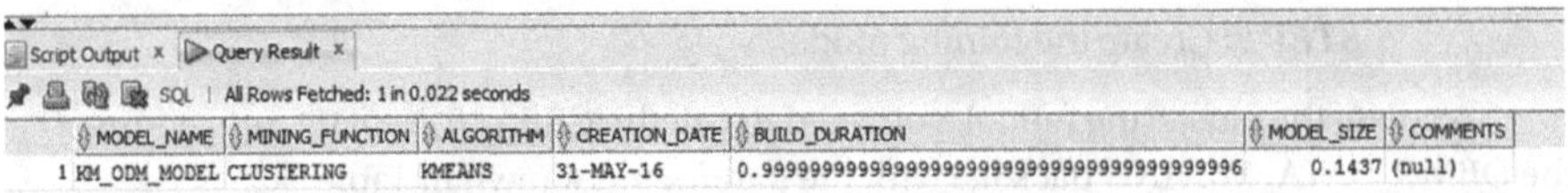

Script Output × | Query Result ×

SQL | All Rows Fetched: 1 in 0.022 seconds

	MODEL_NAME	MINING_FUNCTION	ALGORITHM	CREATION_DATE	BUILD_DURATION	MODEL_SIZE	COMMENTS
1	KM_ODM_MODEL	CLUSTERING	KMEANS	31-MAY-16	0.9999999999999999999999999999999999999996	0.1437	(null)

***Figure 3-7.** User mining model record*

STEP 4: Check the data attributes (Figure 3-8) that were used to create the model

```
SELECT attribute_name, attribute_type
  FROM user_mining_model_attributes
 WHERE model_name = 'KM_ODM_MODEL'
ORDER BY attribute_name;
```

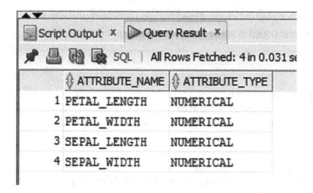

	ATTRIBUTE_NAME	ATTRIBUTE_TYPE
1	PETAL_LENGTH	NUMERICAL
2	PETAL_WIDTH	NUMERICAL
3	SEPAL_LENGTH	NUMERICAL
4	SEPAL_WIDTH	NUMERICAL

Figure 3-8. Model settings for a k-means clustering model

STEP 5: Result evaluation

You need to call function GET_MODEL_DETAILS_KM of DBMS_DATA_MINING package to get model details and rules for each cluster.

```
select DBMS_DATA_MINING.GET_MODEL_DETAILS_KM('KM_ODM_MODEL') from dual
```

View Value

```
SYS.DM_CLUSTERS
SYS.DM_CLUSTER(1,'1',152,NULL,1,4.6296429189750699,NULL,SYS.DM_CHILDREN(SYS.DM_CHILD(2),SYS.DM_CHILD(3)),SYS.DM_CENTROIDS(SYS.
SYS.DM_CLUSTER(2,'2',96,1,2,1.2359570312500101,NULL,SYS.DM_CHILDREN(SYS.DM_CHILD(4),SYS.DM_CHILD(5)),SYS.DM_CENTROIDS(SYS.DM_
SYS.DM_CLUSTER(3,'3',56,1,2,0.74465880102041304,NULL,SYS.DM_CHILDREN(SYS.DM_CHILD(6),SYS.DM_CHILD(7)),SYS.DM_CENTROIDS(SYS.DM_
SYS.DM_CLUSTER(4,'4',36,2,3,0.6229166666666674,NULL,SYS.DM_CHILDREN(SYS.DM_CHILD(12),SYS.DM_CHILD(13)),SYS.DM_CENTROIDS(SYS.DM_
SYS.DM_CLUSTER(5,'5',60,2,3,0.51890555555555395,NULL,SYS.DM_CHILDREN(SYS.DM_CHILD(14),SYS.DM_CHILD(15)),SYS.DM_CENTROIDS(SYS.C
SYS.DM_CLUSTER(6,'6',25,3,3,0.179165680473369,NULL,SYS.DM_CHILDREN(SYS.DM_CHILD(NULL)),SYS.DM_CENTROIDS(SYS.DM_CENTROID('PETA
SYS.DM_CLUSTER(7,'7',31,3,3,0.86424557752340603,NULL,SYS.DM_CHILDREN(SYS.DM_CHILD(8),SYS.DM_CHILD(9)),SYS.DM_CENTROIDS(SYS.DM_
SYS.DM_CLUSTER(8,'8',4,7,4,0.073749999999998594,NULL,SYS.DM_CHILDREN(SYS.DM_CHILD(10),SYS.DM_CHILD(11)),SYS.DM_CENTROIDS(SYS.C
SYS.DM_CLUSTER(9,'9',27,7,4,0.35473251028806602,NULL,SYS.DM_CHILDREN(SYS.DM_CHILD(18),SYS.DM_CHILD(19)),SYS.DM_CENTROIDS(SYS.C
SYS.DM_CLUSTER(10,'10',3,8,5,0.039999999999992,NULL,SYS.DM_CHILDREN(SYS.DM_CHILD(NULL)),SYS.DM_CENTROIDS(SYS.DM_CENTROID('PET
SYS.DM_CLUSTER(11,'11',1,8,5,0,NULL,SYS.DM_CHILDREN(SYS.DM_CHILD(NULL)),SYS.DM_CENTROIDS(SYS.DM_CENTROID('PETAL_WIDTH',NULL,1
SYS.DM_CLUSTER(12,'12',12,4,4,0.38791666666666702,NULL,SYS.DM_CHILDREN(SYS.DM_CHILD(16),SYS.DM_CHILD(17)),SYS.DM_CENTROIDS(SYS
SYS.DM_CLUSTER(13,'13',24,4,4,0.227604166666662,NULL,SYS.DM_CHILDREN(SYS.DM_CHILD(NULL)),SYS.DM_CENTROIDS(SYS.DM_CENTROID('PE
SYS.DM_CLUSTER(14,'14',36,5,4,0.31817901234568302,NULL,SYS.DM_CHILDREN(SYS.DM_CHILD(NULL)),SYS.DM_CENTROIDS(SYS.DM_CENTROID('
SYS.DM_CLUSTER(15,'15',24,5,4,0.21746527777777899,NULL,SYS.DM_CHILDREN(SYS.DM_CHILD(NULL)),SYS.DM_CENTROIDS(SYS.DM_CENTROID('
SYS.DM_CLUSTER(16,'16',3,12,5,0.19777777777778299,NULL,SYS.DM_CHILDREN(SYS.DM_CHILD(NULL)),SYS.DM_CENTROIDS(SYS.DM_CENTROID('
SYS.DM_CLUSTER(17,'17',9,12,5,0.26049382716049901,NULL,SYS.DM_CHILDREN(SYS.DM_CHILD(NULL)),SYS.DM_CENTROIDS(SYS.DM_CENTROID('
SYS.DM_CLUSTER(18,'18',21,9,5,0.0927250000000001501,NULL,SYS.DM_CHILDREN(SYS.DM_CHILD(NULL)),SYS.DM_CENTROIDS(SYS.DM_CENTROID
SYS.DM_CLUSTER(19,'19',6,9,5,0.82388888888888701,NULL,SYS.DM_CHILDREN(SYS.DM_CHILD(NULL)),SYS.DM_CENTROIDS(SYS.DM_CENTROID('P
```

Figure 3-9. Clustering model results

The result is available in the table data type format, which looks a bit messy (see Figure 3-9). We need to use the SQL table type data manipulation functions to convert it into a readable format. The following is an example to view the rules in a readable format for the leaf nodes (see also Figure 3-10).

Note As Oracle prepares *k*-means in a hierarchical fashion, the leaf nodes are the final clusters. Leaf nodes are those nodes where the clustering process stops and doesn't have a child node. The number of leaf nodes is defined by the model-setting parameter: number of clusters. For example, if number of leaf nodes is set to 3, then the final number of clusters is 3.

```
 SELECT model.id rule_id,
        antecedent.attribute_name aname,
        antecedent.conditional_operator operator,
        NVL(antecedent.attribute_str_value,ROUND(antecedent.attribute_num_
        value,4)) value,
        model.rule.rule_support support,
        model.rule.rule_confidence confidence
 FROM TABLE(DBMS_DATA_MINING.GET_MODEL_DETAILS_KM('KM_ODM_MODEL')) model,
       TABLE(model.rule.antecedent) antecedent,
       TABLE(model.child) chl
 where chl.id is null
ORDER BY model.id,support, confidence desc;
```

Script Output × | Query Result ×

SQL | All Rows Fetched: 24 in 0.038 seconds

	RULE_ID	ANAME	OPERATOR	VALUE	SUPPORT	CONFIDENCE
1	3	SEPAL_LENGTH	<=	5.78	42	0.80769230769230804
2	3	PETAL_WIDTH	>=	.1	42	0.80769230769230804
3	3	PETAL_WIDTH	<=	.34	42	0.80769230769230804
4	3	SEPAL_LENGTH	>=	4.19	42	0.80769230769230804
5	3	SEPAL_WIDTH	<=	3.92	42	0.80769230769230804
6	3	SEPAL_WIDTH	>=	2.96	42	0.80769230769230804
7	3	PETAL_LENGTH	<=	2.18	42	0.80769230769230804
8	3	PETAL_LENGTH	>=	1	42	0.80769230769230804
9	4	SEPAL_WIDTH	<=	3.44	34	0.89473684210526305
10	4	PETAL_WIDTH	>=	1.78	34	0.89473684210526305

***Figure 3-10.** Readable rules for k-means cluster rules*

Note GET_MODEL_DETAILS_EM and GET_MODEL_DETAILS_OC are the model details API for Expectation Maximization and O-Cluster clustering models.

Clustering using Oracle R Enterprise

You need to boot up the R client console and follow the following steps for creating clusters using Oracle R Enterprise.

> **STEP 1:** Load the ORE library. This command loads the Oracle R basic functionality packages present in the ORE library to the R environment.

```
library(ORE)
```

> **STEP 2:** Use ore.connect to connect to the database

```
 ore.connect(user = "dmuser", sid = "ORCL", host = "localhost", password =
"sibanjan123",port = 1521)
```

> **STEP 3:** Use ore.sync to synchronize the metadata in the database schema with R environment. In the following code, the ore.sync function creates an ore.frame object that is a proxy to the XXCUST_IRIS_DATA database table.

```
ore.sync("DMUSER","XXCUST_IRIS_DATA",use.keys=TRUE)
```

> **STEP 4:** Use ore.get to get the proxy ore.frame for the table XXCUST_IRIS_DATA and assign it to an R variable.

```
a<-ore.get("XXCUST_IRIS_DATA",schema="DMUSER")
```

> **STEP-5:** The ore.ls function lists all the proxy database objects available for the current environment.

```
ore.ls()
```

> **STEP 6:** Create the KMeans model by invoking the ore. odmKmeans function. It accepts all the parameters and default values that were described in the Table 3-1.

```
km.iris<- ore.odmKMeans(~., a,num.centers=3, auto.data.prep=FALSE)
```

> **STEP 7:** Result evaluation

Use the summary function to see the model summary results (Figure 3-11), which includes the model settings and centroid of the model cluster.

```
summary(km.iris)
```

```
R  R Console                                                          [_][□][x]

 >   summary(km.iris)

Call:
ore.odmKMeans(formula = ~., data = a, auto.data.prep = FALSE,
    num.centers = 3)

Settings:
                             value
clus.num.clusters               3
block.growth                    2
conv.tolerance               0.01
distance                euclidean
iterations                      3
min.pct.attr.support          0.1
num.bins                       10
split.criterion          variance
prep.auto                     off

Centers:
      ID SEPAL_LENGTH SEPAL_WIDTH PETAL_LENGTH PETAL_WIDTH centersB[, -1L]
2 123.0     6.383051    2.964407     5.218644    1.8779661  Iris-virginica
4  68.5     5.854000    2.874000     3.930000    1.1980000 Iris-versicolor
5  22.0     5.009302    3.409302     1.451163    0.2348837     Iris-setosa

 >  |
```

Figure 3-11. Model results

Every cluster formed has some common characteristics that are defined as rules by the clustering algorithms to uniquely identify each cluster. To view these rules, you can use the R rules function as shown in Figure 3-12.

```
rules(km.iris)
```

```
R  R Console                                                          [_][□][x]

    rhs.cluster.id rhs.support rhs.conf lhs.support  lhs.conf        lhs.var
1                1         152        1         136 0.8947368          CLASS
2                1         152        1         136 0.8947368          CLASS
3                1         152        1         136 0.8947368          CLASS
5                1         152        1         136 0.8947368             ID
4                1         152        1         136 0.8947368             ID
7                1         152        1         136 0.8947368   PETAL_LENGTH
6                1         152        1         136 0.8947368   PETAL_LENGTH
9                1         152        1         136 0.8947368    PETAL_WIDTH
8                1         152        1         136 0.8947368    PETAL_WIDTH
11               1         152        1         136 0.8947368   SEPAL_LENGTH
10               1         152        1         136 0.8947368   SEPAL_LENGTH
13               1         152        1         136 0.8947368    SEPAL_WIDTH
12               1         152        1         136 0.8947368    SEPAL_WIDTH
    lhs.var.support lhs.var.conf           predicate
1               151    0.2500000      IN Iris-setosa
2               151    0.2500000  IN Iris-versicolor
3               151    0.2500000   IN Iris-virginica
5               152    0.0000000           <= 152
4               152    0.0000000            >= 1
7               147    0.3333333           <= 6.31
6               147    0.3333333            >= 1
9               152    0.3000000           <= 2.5
8               152    0.3000000            >= 0.1
11              144    0.2500000           <= 7.37
```

Figure 3-12. Cluster rules detail

The results from this model can also be retrieved from the SQL environment using DBMS_DATA_MINING.GET_MODEL_DETAILS_KM function that I described in an earlier section. The ORE model gets stored with an ORE specific internal name (Figure 3-13). This name is required to be passed as a parameter to the model details PLSQL API to retrieve the results in an SQL environment. The internal name can be found by executing the following command in the R environment.

```
km.iris$name
```

```
> km.iris$name
[1] "ORE$18_20"
>
```

Figure 3-13. *Internal ORE model name*

You can use the preceding extracted name in the model details PLSQL API to retrieve the results in the Oracle SQL environment (Figure 3-14).

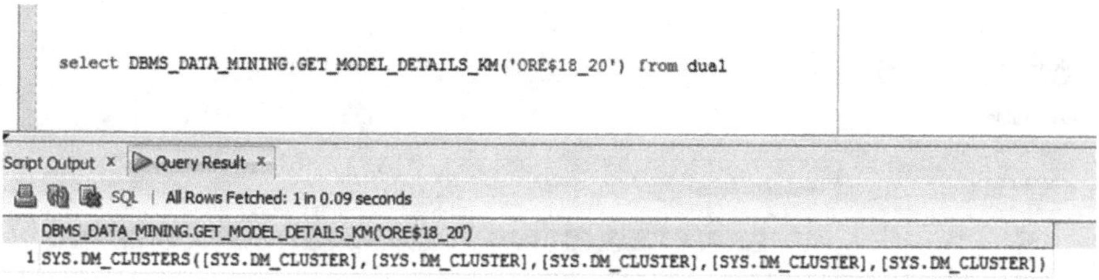

```
select DBMS_DATA_MINING.GET_MODEL_DETAILS_KM('ORE$18_20') from dual
```

Script Output × ▷ Query Result ×
🖳 🐼 🖳 SQL | All Rows Fetched: 1 in 0.09 seconds

DBMS_DATA_MINING.GET_MODEL_DETAILS_KM('ORE$18_20')
1 SYS.DM_CLUSTERS([SYS.DM_CLUSTER],[SYS.DM_CLUSTER],[SYS.DM_CLUSTER],[SYS.DM_CLUSTER],[SYS.DM_CLUSTER])

Figure 3-14. *ORE model results in the SQL environment*

> **STEP 8:** Sometimes it is necessary to keep the model in a data store to reuse it anytime later. It is an optional step but is recommended for the model reusability.

```
ore.save(km.iris, name="km_ore_cluster_iris",overwrite=TRUE)
```

Creating a cluster model using SQL Developer

A clustering model can be created from the data mining option in SQL developer. This needs no coding and is suitable for business analysts who are not acquainted with programming. Refer to the steps discussed in Chapter 2 to create a new project and workflow. Once a new workflow is created, follow the steps given here to create a cluster model in SQL Developer.

STEP 1: Define a data source to be used for creating a clustering model (Figure 3-15). Drag the data source node from the data section of the components panel to the worksheet.

Figure 3-15. *Data Source node in ODM workflow*

STEP 2: Select the view DMUSER.XXCUST_IRIS_DATA_V for the data source as shown in Figure 3-16.

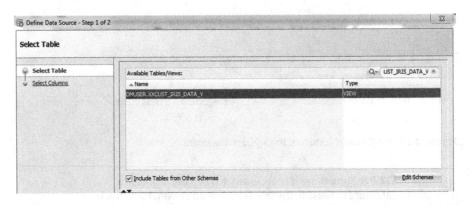

Figure 3-16. *Define Data Source*

STEP 3: Select and drag clustering to the worksheet from the Model section in Components editor. Connect the Data Source node and Clustering node as shown in Figure 3-17.

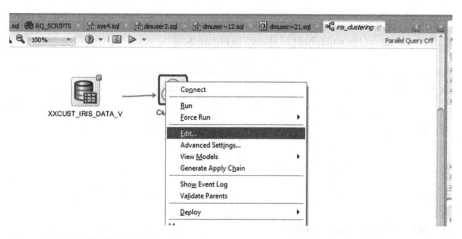

Figure 3-17. *Connect XXCUST_IRIS_DATA_V and Clust Build node*

STEP 4: For a new cluster node, all three available algorithms get defaulted for execution. As we want to perform only *k*-means clustering, we can remove the other two algorithms from the list. Right-click the cluster node and select Edit to change the model settings (see Figure 3-18).

Figure 3-18. *Edit Clust Build node*

STEP 5: From the algorithm section, select O-Cluster and press the delete option as shown in Figure 3-19. Repeat the same for the Expectation Maximization algorithm.

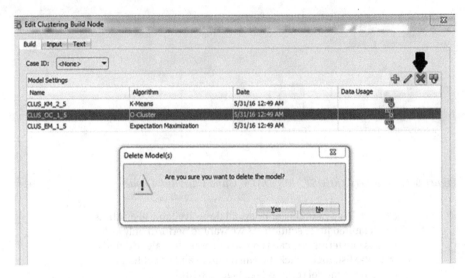

Figure 3-19. *Remove models except K-Means algorithm*

> **STEP 6:** Run the workflow by pressing the play button at the top of the worksheet as shown in Figure 3-20.

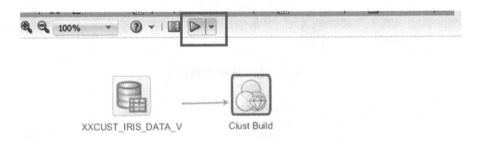

Figure 3-20. *Execute the workflow*

> **STEP 7:** Result evaluation

To view the results, right-click on the Clust Build node and select the model from the view model option. Once the model result appears on the screen, clusters are shown in a hierarchy (see Figure 3-21). Click a node of the tree to see the corresponding rule for that cluster, which appears at the rule section at the bottom of the screen. You can also view the centroid of the currently selected cluster. A centroid is the most probable value for that cluster.

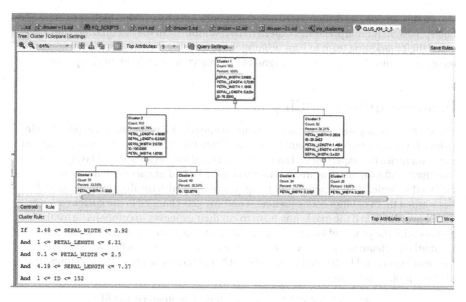

Figure 3-21. *Clusters hierarchy*

■ **Note**   In the following case study, you will see the methods to interpret, validate, and store the result of a cluster model in the database.

Case Study—Customer Segmentation

For a company to be successful, it is important to understand their customer's needs and not exhaust their budget by advertising their products to each customer. Every customer is different but has one or more characteristics in common that they exhibit for the product needs. Companies need to find these hidden attributes and design custom programs that are unique to each group. This process to divide customers into similar and identifiable segments is known as customer segmentation. The consumers find tailored promotional events relevant to their needs, and it encourages them to tap this opportunity. As a by-product, personalized promotions are highly valued by them, and this leads to better customer loyalty. For organizations, this earns higher returns than targeting each and every customer. Also, organizations need to be cost-effective and target those customers where they can increase their revenue. Adding more to the costs that bring down the profit doesn't serve the purpose of a promotional activity until and unless there is some tertiary benefit attached to it. Organizations need to prioritize their customers to derive value out of them. They need to design an efficient segmentation process that overlaps internal constraints and customer's needs. Also, it is essential that the analytical CRM (Customer Relationship Management) system should automatically assign a segment to a new customer when they are acquired. This helps to avoid the recurring costs of doing the customer segmentation exercise each and every time.

93

In this section, we design a customer segmentation process for a high-tech store Sant Enterprises. Sant Enterprises has experienced low footfalls in the last quarter. They have asked their marketing team to start promotions to increase their store's footfall and also increase revenue. You have been assigned as an analytics consultant for this project.

Business understanding

You start discussing with the marketing team to understand what Sant Enterprises' pain points are and the target they want to achieve. From the discussion, you understand that Sant Enterprise has experienced fewer customers and lost revenue in the last quarter. They have had a dip in their profits, and this has had a bad impact in their quarterly financial statement. The management is quite unhappy with the marketing team and has warned them of bitter consequences if nothing gets changed in the current quarter. They have also asked the marketing team to cut their expenses to compensate for the loss and stabilize the year-end financial report. Based on the business scenario, you decide the marketing team should target their existing GOLD customers who had been their frequent buyers and high revenue earners but have recently not visited their stores. You plan the project as follows:

- Segment the customers by a value-based approach using RFM (recency, frequency, monetary) methodology. It is a method to analyze customer value by computing a score for each customer. The scores help identify high-response customers who can contribute to improving the overall response rate of a marketing campaign.

- Filter the GOLD customers based on their RFM score.

- Further segment these GOLD customers to cluster them into homogeneous groups based on their profile and purchase history.

- Understand the important customer attributes and rules that define each cluster. This would provide the marketing team with inputs for personalized promotional offers.

- Store this result back to the operational database for the marketing team to integrate with their CRM system.

- Implement real-time segment assignment when a new customer arrives.

Data understanding

The marketing team liked your approach of solving the problem and provided access to their CRM database. You need to kick off the project by understanding the metadata of the CRM database (see Table 3-3):

Table 3-3. MetaData of the Tables in the CRM Database

Table	Description
CUSTOMERS	Contains important customer profile information
SALES	Has historical sales transaction data for each customer
PRODUCTS	The details of all the products of SANT Enterprises
PROMOTIONS	Details of past promotional activities

There are some columns in the table that are not required. You require the information that influences SANT Enterprise's sales and customer's purchasing behavior such as

- Demographic information (age, gender, income, etc.)

- Psychographic (lifestyle) information

- Behavioral information (such as spending and purchase history)

Based on this knowledge, you create a database view CUST_TRANSACTION_V that provides you with only the previously mentioned data from the SANT Enterprises CRM database.

RFM Segmentation—Data Preparation

First, you need to create a new data mining project and start defining the data sources. Follow the steps outlined in Chapter 2 to create a new project and a data mining workflow. The project and workflow names are as in Figure 3-22.

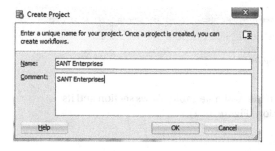

Figure 3-22. Create ODM project and workflow

95

For calculating the RFM score for customers, you need the following data that are available in the sales table of SH schema of the Oracle Database.

- Revenue for each customer

- Last purchase date

- Number of visits to the store by a customer

Define a sales data source by dragging the data source node from the data section of the components panel to the workflow editor. While defining the data source, if the sales table doesn't appear, select "Include tables from other schema," and then select SH schema as shown in Figure 3-23.

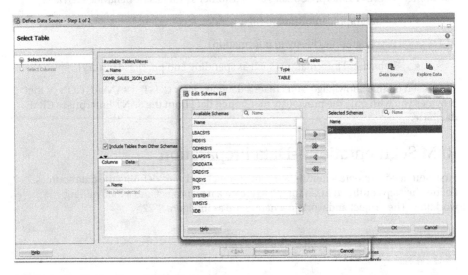

Figure 3-23. *Define Data Source dialog box*

Confirm the sales table appears in the Available Table/Views section and its attributes appear in the Columns section (Figure 3-24).

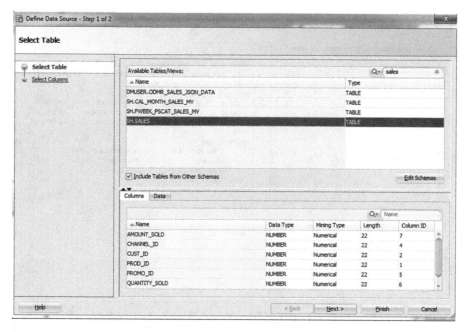

Figure 3-24. Select Sales Data table

RFM Segmentation—DATA Modeling

Drag an SQL Query node from the data section to the workflow area. Connect the Sales data source to the SQL Query (Figure 3-25).

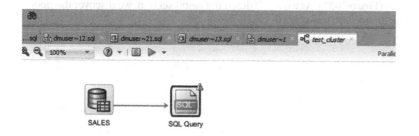

Figure 3-25. Connect Sales node to SQL Query node

Right-click on the SQL Query node and select Edit to include SQL query for adding the RFM SQL statement as shown in Figure 3-26. The SQL statement calculates an RFM score that places each customer into certain deciles. For example, the top 20% of customers, based on their recent purchases are placed into recency decile 5; the next 20% into recency decile 4, and so on. Similarly, the top 20% of customers, based on their frequency and total purchase amount, are placed into frequency and monetary decile 5, respectively. These scores are then added to calculate an overall score for each customer.

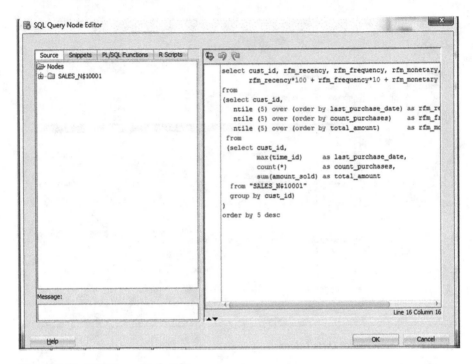

Figure 3-26. *SQL Query Node Editor*

Figures 3-27 and 3-28 show the snapshots of the RFM score output after executing the SQL query node. The best customers who ranked excellent in the three categories (recency, frequency, and monetary) are assigned a score of 555 and are displayed at the top of the results (Figure 3-27). As we scroll down the result section, we discover the poor-performing customers with low scores.

	CUST_ID	RFM_RECENCY	RFM_FREQUENCY	RFM_MONETARY	RFM_COMBINED
1	13,113	5	5	5	555
2	11,108	5	5	5	555
3	4,713	5	5	5	555
4	1,593	5	5	5	555
5	181	5	5	5	555
6	7,993	5	5	5	555
7	2,448	5	5	5	555
8	9,502	5	5	5	555
9	9,213	5	5	5	555
10	2,656	5	5	5	555
11	6,013	5	5	5	555
12	14,270	5	5	5	555

Figure 3-27. *RFM scores*

	CUST_ID	RFM_RECENCY	RFM_FREQUENCY	RFM_MONETARY	RFM_COMBINED
721	1,688	5	4	3	543
722	3,177	5	4	3	543
723	9,168	5	4	3	543
724	7,517	5	4	2	542
725	3,182	5	4	2	542
726	4,152	5	4	2	542
727	817	5	3	5	535
728	7,748	5	3	5	535
729	4,845	5	3	5	535
730	210	5	3	5	535
731	10,373	5	3	5	535
732	3,643	5	3	5	535
733	1,626	5	3	5	535
734	9,881	5	3	5	535

Figure 3-28. *Poor-performing customers*

99

Need-Based Segmentation—Data Preparation

As we are interested in creating need-based segmentation for only GOLD standard customers, we will filter those customers from the rest. To do that, drag a Filter Rows node from the Transforms section to the workflow worksheet (Figure 3-29).

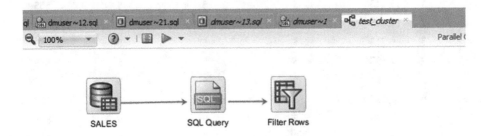

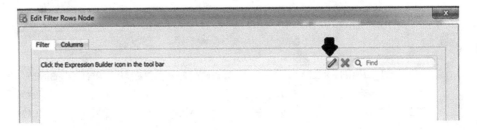

Figure 3-29. *Connecting SQL Query node with Filter Rows node*

Right-click on the Filter Rows node and select edit. This results in a pop up of an expression worksheet as shown in the Figure 3-30. To add a filter, click on the pencil icon to build the expression for the filter condition.

Figure 3-30. *Build expression for Filter condition*

In the expression builder field, write the expression for the filtering condition and click on the validate button to validate correctness of the syntax (Figure 3-31).

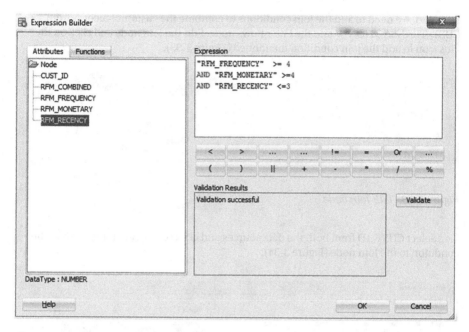

Figure 3-31. Expression builder for filter conditions

Next, we need to join this filtered RFM data with the customer profile attributes for creating value-based clusters. To do that, create a data source for the view CUST_TRANSACTION_V that was built during the data understanding phase. Then, drag the Join node from the Transforms section to the workflow editor. Connect the CUST_TRANSACTION_V view and the Filter Rows node to the Join node as shown in Figure 3-32.

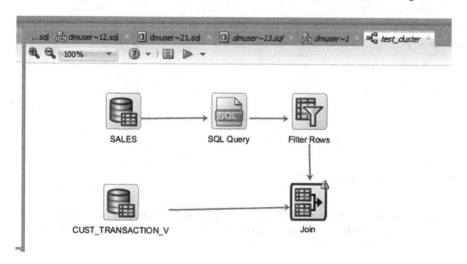

Figure 3-32. Connect CUST_TRANSACTION_V and Filter Rows node to the Join node

Next, we need to add the join conditions to combine the customer records from CUST_TRANSACTION_V and the Filter Rows node. Edit the Join node and click on the Plus icon to add the join condition mentioned (Figure 3-33).

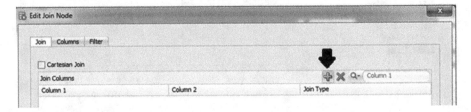

Figure 3-33. *Edit Join node*

Select CUST_ID from both the data sources and click on the <u>A</u>dd button to add the condition to the Join node (Figure 3-34).

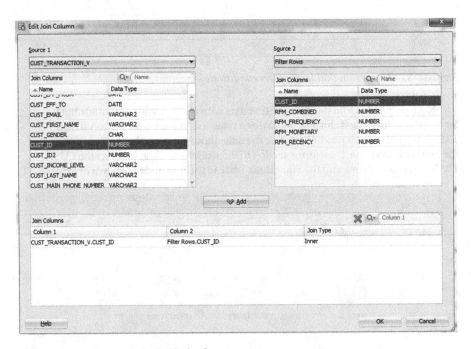

Figure 3-34. *Edit Join Column dialog box*

Next, we need a Transform node to view the joined data and carry out required data transformations. Drag a Data Transformation node from Transforms panel to the workflow editor (Figure 3-35).

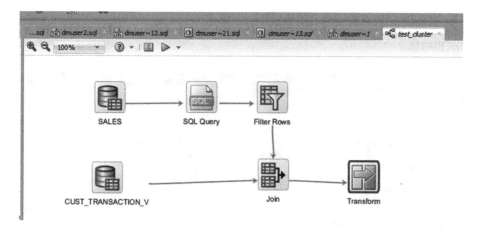

Figure 3-35. *Connect Transform node to Join node*

Right-click the transform node and select edit to view the table columns. For this data set, there are some columns such as ID columns that don't add any value to the model. Also, some variables such as CUST_YEAR_OF_BIRTH and RFM_COMBINED are assigned number data type automatically by the system. However, if we examine the data for this variable, it is discrete in nature and so should be of character data type. The values for discrete data are distinct in nature and so are recognized as categorical data types. Continuous data is indicated as a number data type, as it is not restricted to a separate value and occupies any value over a continuous range.

Exclude the following columns (Table 3-4) by clicking on the Output Column next to the column name, as these are not to be used for building the clustering model.

Table 3-4. *Excluded Columns*

Excluded Columns	
CHANNEL_ID	PROD_SUBCATEGORY_ID
CUST_ID	PROD_TOTAL_ID
CUST_ID1	PROD_CATEGORY_ID
CUST_ID2	PROMO_ID
PROD_CATEGORY_DESC	PROMO_ID1
PROD_CATEGORY_ID	PROMO_TOTAL_ID
PROD_DESC	RFM_COMBINED
PROD_ID	RFM_FREQUENCY
PROD_ID1	RFM_MONETARY
PROD_SRC_ID	RFM_RECENCY
PROD_SUBCATEGORY_DESC	SUPPLIER_ID

Next, we change the datatype of CUST_YEAR_OF_BIRTH and RFM_COMBINED. Click on the plus icon as shown in Figure 3-36 to change the data type of CUST_YEAR_OF_BIRTH.

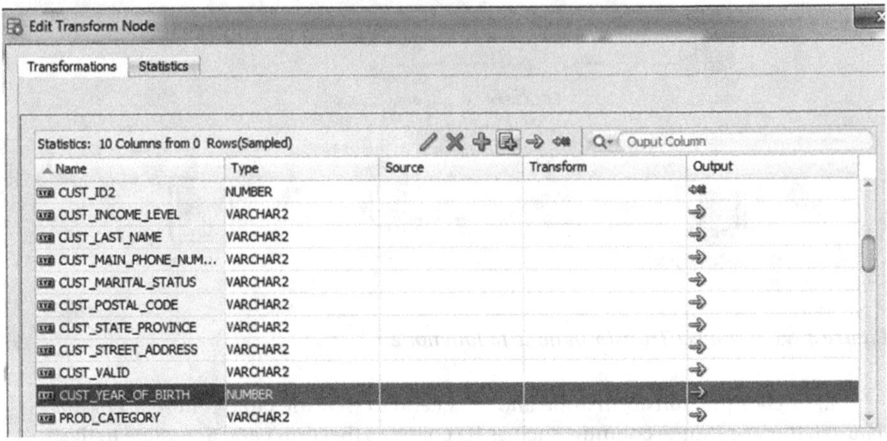

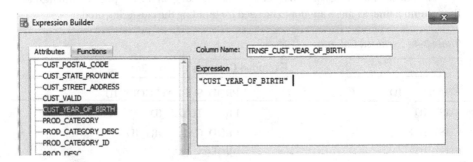

Figure 3-36. *Edit Transform Node dialog box*

In the Expression Builder, enter the name of the column as "TRNSF_CUST_YEAR_OF_BIRTH" to indicate it as a transformed variable of CUST_YEAR_OF_BIRTH and select CUST_YEAR_OF_BIRTH from the attribute list (Figure 3-37).

Figure 3-37. *Expression Builder for the transform node*

In the Functions tab, expand the conversion folder and select the TO_CHAR function for number to character data type conversion. The final expression should appear as shown in Figure 3-38.

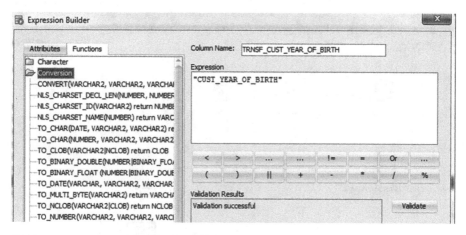

Figure 3-38. *Expression Builder for transformations*

Carry out the same process to transform the data type of attribute RFM_COMBINED from NUMBER to VARCHAR2, which is a character data type (Figure 3-39).

Figure 3-39. *Transformed variables*

As this segmentation exercise is to help the marketing team to target only currently active GOLD customers, we need to select only the active customers from our database. To do that, create another Filter node in the workflow sheet (Figure 3-40).

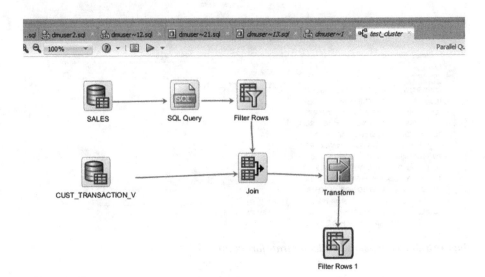

Figure 3-40. *Filter Rows 1 connected to Transform node*

Edit the filter node and build an expression to select only active customers as shown in Figure 3-41.

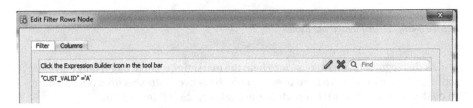

Figure 3-41. *Edit Filter Rows Node dialog box*

Need-Based Segmentation—Data Modeling

Drag the clustering node from Models section and connect it to the Filter Rows 1 node. From the properties panel, change the name of the node to sant_clusters for its better visibility and to distinguish it from other nodes (Figure 3-42).

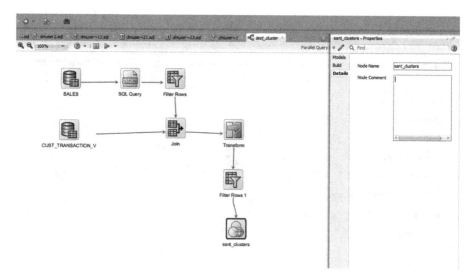

Figure 3-42. *Node rename to sant_clusters*

Edit the clustering node, and select only the "K-Means" algorithm.

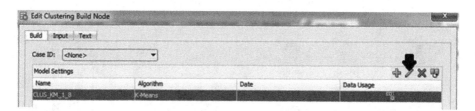

Figure 3-43. *Edit Clustering Build Node dialog box*

Click on the pencil icon in the model settings row to edit the Algorithm properties as shown in Figure 3-43. The model settings user interface shows two tabs—Data Usage and Algorithm Settings (Figure 3-44).

- The Data Usage tab has the inputs that are to be used for building the model.

- The Algorithm Settings tab shows the setting parameters for the model.

Based on the marketing team's advice, we need to create three clusters. Navigate to the Algorithm tab and change the setting Number of Clusters to 3 (Figure 3-44).

Data Usage	Algorithm Settings

The default settings should work well for most use cases. For information on changing model algorithm settings, click Help.

Number of Clusters:	3
Growth Factor:	2
Convergence Tolerance:	0.01
Distance Function:	Euclidean ▼
Number of Iterations:	30
Min Percent Attribute Rule Support:	0.1
Number of Histogram Bins:	10
Split Criterion:	Variance ▼

Figure 3-44. K-Means Algorithm Settings

Execute the node; and once it completes successfully, view the results by selecting the model from the View Models option as shown in Figure 3-45.

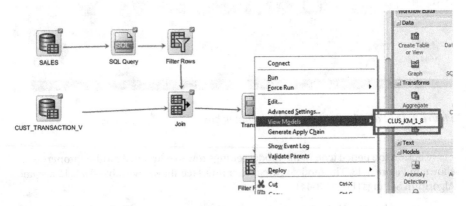

Figure 3-45. View Models option

Result Evaluation

You can see a tree structure for the clusters built (Figure 3-46). The leaf nodes are the three desired clusters.

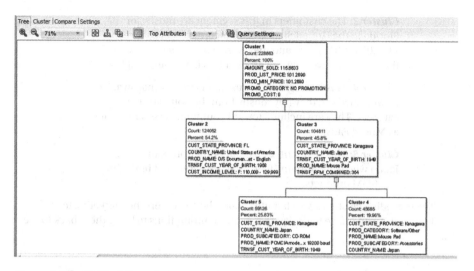

Figure 3-46. Cluster hierarchy

Understanding the clusters

Before finalizing the clusters, it is important to understand the clusters created by the algorithm. The cluster analysis is not very useful if it doesn't give enough information to categorize the clusters. The cluster results should give some unique information related to customers that can define the segments.

If we analyze the three final clusters for high value but recently inactive customers of Sant Enterprises, we find out some of the customer attributes that uniquely define each segment (Figure 3-47).

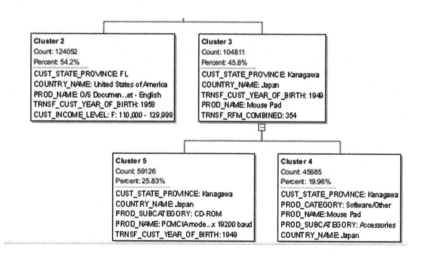

Figure 3-47. Three-leaf node Clusters 2, 5, and 4

Cluster 2: The customers in this segment are mostly for Florida in the United States. Their income level is between $110,000 to $129,999, and they were born in 1958. They were the frequent buyers for the product O/S Document-English.

Cluster 4: This is a segment of customers from Kanagawa, Japan. However, they buy mostly from the Software/Other category. They are inclined toward computer accessories such as Mouse Pads.

Cluster 5: This segment of customers are from Kanagawa, Japan who were born in 1949. They are a frequent buyers of CD-ROM products.

The preceding clusters give a fair idea about the customers the marketing team for Sant Enterprises can utilize to design personalized promotions to bring them back to the store.

Deployment—Storing the Results Back to the Database

The insights are useful when they are put into action and stored in the operational database for future reference. Oracle Advanced Analytics sits over the Oracle Database, so integrating the results back to operations in real time is faster than using other data science tools.

To store the results back to the database, drag an Apply node from the "Evaluate and Apply" section of the component panel. Connect this node to the sant_clusters node and the CUST_TRANSACTION_V data source as shown in Figure 3-48.

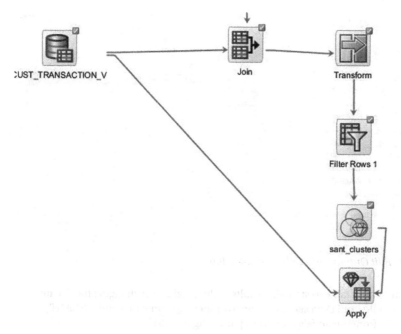

Figure 3-48. Connect Apply node to the sant_clusters node and CUST_TRANSACTION_V node

Edit the Apply node, click on the plus icon to add CUST_ID as an additional column, and navigate to the Additional Output column as shown in Figure 3-49.

Figure 3-49. Edit Apply Node dialog

Once you select CUST_ID as an additional attribute to be a part of the stored cluster results, select OK to save the details (Figure 3-50). This additional information for the result set provides a key to joining the customer profile information with the clustering results.

111

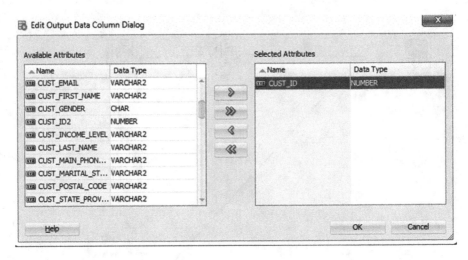

Figure 3-50. Edit Output Data Column Dialog box

Create an output table to store the results in the database by dragging the Create Table or View node from Data section to the worksheet. Rename this node to SANT_CLUS_OUTPUT from the node's properties panel (Figure 3-51).

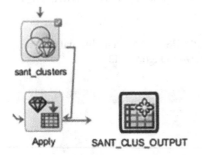

Figure 3-51. Output Table node renamed to SANT_CLUST_OUTPUT

To define the table for storing the result, right-click on the SANT_CLUS_OUTPUT node and select edit. Enter CUST_CLUSTER_RESULTS as a name for the new table. Click OK to return to the workflow editor (Figure 3-52).

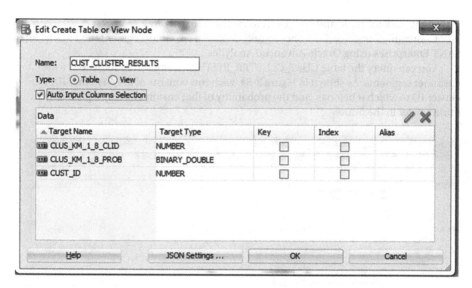

Figure 3-52. Edit Create Table or View Node dialog box

The final workflow looks like Figure 3-53.

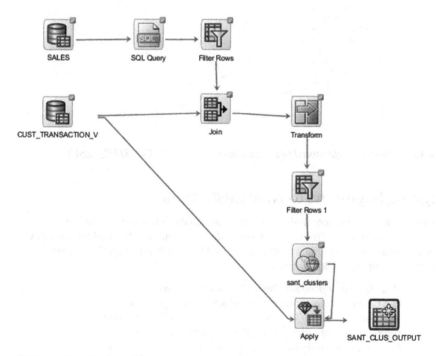

Figure 3-53. Final workflow

Right-click the node and select Run to execute the workflow.

Congratulations!! You have successfully created a customer segmentation model for SANT Enterprises using Oracle Advanced Analytics.

You can query the table CUST_CLUSTER_RESULTS from the database to see the customer segments. As shown in Figure 3-54, each row consists of the customer ID, the cluster ID to which it belongs, and the probability of that customer being a part of the same cluster in the future.

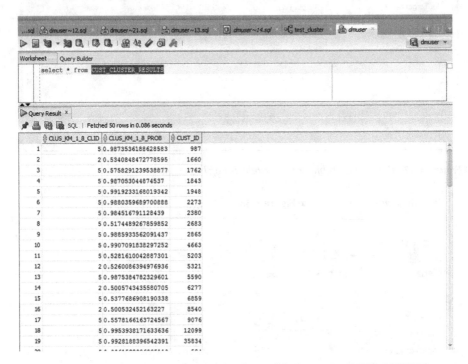

Figure 3-54. *Cluster results stored in the database table CUST_CLUSTER_RESULTS*

Assigning Segments to New Customers

When new customers are acquired, it is better to assign them to a segment at the beginning of their journey. This will help the marketing team to be focused and proactive in understanding their customers from day one. In Oracle Advanced Analytics, new cluster assignment can be done in two ways.

> ***Real-time assignment:*** The real-time cluster assignment can be done by invoking cluster_id and cluster_probability functions from an SQL query as shown in the following code segment. You can make this SQL statement a part of the CRM application that gets triggered when a new customer is entered into the system.

```
SELECT cust_id,
       cluster_id(CLUS_KM_1_8 USING *) as Cluster_Num,
        cluster_probability(CLUS_KM_1_8, cluster_id(CLUS_KM_1_8 USING *)
USING *) as Cluster_Prob
FROM   CUST_TRANSACTION_V
```

> **Batch process:** In a fast-moving business environment
> such as a supermarket, it is not always necessary to process
> customer segments in real time. There is not much business
> impact if the customers are assigned to segments in an
> overnight batch process or during weekends.

You can use the APPLY procedure in DBMS_DATA_MINING package for a batch apply operation, and this can also be automated to run at a certain schedule using DBMS_SCHEDULER.

```
SET SERVEROUTPUT ON;
BEGIN
dbms_data_mining.apply
       (model_name => 'CLUS_KM_1_8',
        data_table_name => 'CUST_TRANSACTION_NEW',
        case_id_column_name => 'cust_id',
        result_table_name => 'CUST_SEGMENT_RESULTS');
END;
/
```

Summary

The purpose of this chapter was to introduce you to the different clustering methods that Oracle Advanced Analytics offers and the process through which it can be automated for a particular business case. Clustering plays an important role in data science where the data attributes can be grouped into homogeneous segments. We interactively studied the k-means algorithm in detail and the various ways through which this algorithm can be applied from different platforms. The other two clustering algorithms present in Oracle Advanced Analytics algorithms can be used in a similar fashion with only changes to their model settings. The OTN docs on Oracle Data Mining have a detailed list of parameters required for each algorithm. Also, not all clustering algorithms are suitable for each and every data set. Certain criteria such as the number of attributes and records in the data set, the data type of the attributes, and the number of outliers can be used as a deciding factor in choosing a clustering algorithm.

CHAPTER 4

■ ■ ■

Association Rules

Almost anyone who steps into the field of data science would have certainly heard the famous diaper and beer story. An analytical study conducted by a major grocery store on their transactional purchase history found that a man between 30 and 40 years of age, shopping between 5 p.m. to 7 p.m. on Fridays who buys diapers were most likely to have beers in their cart. This motivated the grocery store to have the beer isle close to the diaper isle, and the result was a boom of about 35% increase in sales of both the products. This was a fascinating pattern that they found out from data that was invisible to the naked eye and the experience of the store management. So, how did the store find out such an unusual pattern and how do you find such interesting rules for your business? The opportunity lies in conducting an affinity analysis on your data. Affinity analysis implements association rule algorithms at its core that are used to find frequent events that co-occur. Some of its possible uses in various industries include

- Market Basket Analysis: Used by retailers to identify products that have a high probability of being purchased together

- Recommendation engine: Used by both online and offline (brick-and-mortar) stores to recommend products to customers

- Manufacturing organizations: Can be used to find associations and correlations between different activities that affect the quality of a product

- Health care: Used to diagnose possible medical symptoms based on a patient's history

- Internet of Things: Used to have an insight on related events that cause machine failure

- Insurance: Used to know customers preferences on insurance products such as health insurance and motor insurance purchased together. This information can be used to create product bundles or for upselling and cross-sell promotion campaigns.

© Sibanjan Das 2016
S. Das, *Data Science Using Oracle Data Miner and Oracle R Enterprise*,
DOI 10.1007/978-1-4842-2614-8_4

In this chapter, I discuss the following topics:

- Introduction to association rules
- Terminologies associated with association rules
- Fundamentals of Apriori algorithms
- Identification of interesting rules
- Association rules using Oracle SQL and PLSQL APIs
- Association rules using Oracle R Enterprise
- Association rules using Oracle SQL Developer GUI
- Case Study: Market Basket Analysis

Introduction to Association Rules

Association rules help to discover the relationship between different activities that co-occur together. Given a set of transactions, association rules are useful in identifying patterns that uncover relationships and connections among the rules' various attributes. A proper association rule is in the form of "if-then" statements and are formed as follows:

{ Item A} -> {Item B} with a support of X% and Y% confidence

The preceding rule implies that there is a relationship between Item A and Item B such that whenever Item A occurs, Item B also occurs. Here, Item A is known as an antecedent and Item B as a consequent. The support and confidence are the metrics that helps to identify the rule strength. I will be discussing the metrics in a later section.

Terminologies Associated with Association Rules

There are certain terminologies specific to association rule analysis that are necessary to understand prior to discussing the working of association rules. This terminology style closely resembles the language used in retail, as it initially originated to conduct basket analysis for customer transactions.

Transaction: This is the record that contains information about the occurrence of certain events. For example, a purchase transaction records the details of a purchase made by a customer. It might include the details of the item purchased, customer name and address, the sales amount, and so forth. A typical purchase transaction would look like Figure 4-1:

Transaction ID	Transactio n Date	Customer ID	Item Purchased	Unit Sale Price(In Dollars)
T01	11/06/2016	C01	Bread	1.2
T01	11/06/2016	C01	Butter	2
T02	11/06/2016	C02	Milk	3.5
T02	11/06/2016	C02	Bread	1.2
T02	11/06/2016	C02	Egg	1.2
T03	11/06/2016	C03	Bread	1.2
T03	11/06/2016	C03	Butter	2
T04	12/06/2016	C01	Cornflakes	2
T04	12/06/2016	C01	Milk	3.5
T04	12/06/2016	C01	Butter	2
T04	12/06/2016	C01	Egg	1.2
T05	12/06/2016	C03	Bread	1.2
T05	12/06/2016	C03	Milk	3.5
T05	12/06/2016	C03	Egg	1.2

Figure 4-1. *Example purchase transactions data*

K-Item sets: These are collections of *K* items such as {bread, butter}. The set {bread, butter} is a two-item set. Individual elements {Bread} and {Butter} are considered a one-item set each.

Support: This is the frequency of occurrence of particular item sets over the total number of transactions. For example, if the transactions with an item set {Bread, Butter} occur 3 times out of 5 transactions that are analyzed, the support is 0.6 or 60%.

Support = Frequency(rule)/Number of transactions

Confidence: Confidence is the ratio of a rule's support to that of the number of transactions that includes only the antecedent of the transaction. For example, bread had been purchased 4 times out of 5 transactions, which has the support of 0.8. Out of those, bread and butter are bought together three times, with a support of 0.6. Therefore, the confidence of butter to be put along with bread is 0.6/0.8 = 0.75 or 75%.

Confidence = Support(rule)/Support(Antecedent)

Association rule algorithms are computationally resource intensive, as they recursively search for *k* different items to discover item sets and rules; so a classic algorithm "Apriori" is widely used to create association rules. The key idea and assumption in the Apriori algorithm is monotonicity, that is, a subset of a frequent item set is also an item set. This helps reduce the number of computational passes over the data to find the rules. A minimum support and confidence threshold is specified before the execution of the algorithm. The goal is to find item sets that are above the specified minimum support and confidence threshold.

Oracle Advanced Analytics implements the Apriori algorithm for association rules.

Working of an Apriori Algorithm

Apriori employs an iterative method to find all frequent item sets and then generate rules out of those frequent item sets. It reduces the number of candidates being explored by examining only those item sets whose support is greater than the minimum support count (see Figure 4-2). All infrequent item sets are pruned and are not used for rule generation.

Minimum Support	0.4
Minimum Confidence	0.6
Maximum Rule Length	2

Figure 4-2. *Minimum setting parameters*

> *STEP 1:* Start with one-item sets and calculate support of each item (see Figure 4-3).

Item	Count	Support
Bread	4	0.8
Butter	3	0.6
Milk	3	0.6
Egg	3	0.6
Cornflakes	1	0.2

Figure 4-3. *Support calculation*

> *STEP 2:* Prune the items with less than a minimum support threshold of 0.4 and retain the remaining items for further analysis (see Figure 4-4).

Item	Count	Support
Bread	4	0.8
Butter	3	0.6
Milk	3	0.6
Egg	3	0.6

Figure 4-4. *Items above the minimum support threshold*

> *STEP 3:* Generate all two-item sets for the transactional records (Figure 4-5).

	Itemset
T01	(Bread, Butter)
T02	(Milk,Bread),(Milk, Egg),(Bread,Egg)
T03	(Bread, Butter)
T04	(Cornflakes,Milk),(Cornflakes,Butter),(Cornflakes,Butter),(Milk,Butter),(Milk,Egg),(Butter,Egg)
T05	(Bread,Milk),(Bread, Egg),(Milk,Egg)

Figure 4-5. *Two-item sets*

STEP 4: Calculate the support of each two-item set (Figure 4-6).

	Support
(Bread, Butter)	0.4
(Milk,Bread)	0.6
(Milk, Egg)	0.6
(Bread,Egg)	0.6
(Cornflakes,Milk)	0.2
(Cornflakes,Bread)	0.2
(Cornflakes,Egg)	0.2

Figure 4-6. *Support for each two-item set*

STEP 5: Find the frequent two-item sets by considering only those item sets that have support more than the minimum support threshold (Figure 4-7).

	Support
(Bread, Butter)	0.4
(Milk,Bread)	0.4
(Milk, Egg)	0.6
(Bread,Egg)	0.6

Figure 4-7. *Frequent two-item sets*

STEP 6: Calculate the confidence of each rule in the considered frequent two-item set (Figure 4-8).

Itemset	Rules	Antecedent Support	Confidence
(Bread, Butter)	Bread -> Butter	0.8	0.5
	Butter -> Bread	0.4	1
(Milk,Bread)	Milk -> Bread	0.6	0.666666667
	Bread -> Milk	0.8	0.5
(Milk, Egg)	Milk -> Egg	0.6	1
	Egg -> Milk	0.6	1
(Bread,Egg)	Bread -> Egg	0.8	0.5
	Egg -> Bread	0.6	0.666666667

Figure 4-8. *Confidence calculation*

STEP 7: Consider only those rules above the minimum confidence as the final set of association rules (Figure 4-9).

Association Rules	Support	Confidence
Butter -> Bread	0.4	1
Milk -> Bread	0.4	0.666666667
Milk -> Egg	0.6	1
Egg -> Milk	0.6	1
Egg -> Bread	0.4	0.666666667

Figure 4-9. *Two-item sets above the minimum confidence threshold*

Identify Interesting Rules

When the number of items is large, a huge number of association rules are produced. So the task of finding out the relevant and interesting insights out of the generated rules is a cornerstone of a data science project involving association rule mining. Interestingness of rules involves examining them using the functional domain knowledge as well as statistical measures. For example, a store manager, based on his experience, might not want a rule with bread and butter as an insight, although the statistical measures for the same rules are high.

The primary statistical measures for association rules are support, confidence, and lift ratios. As discussed earlier, support and confidence form the building blocks to create the association rules. A rule with high confidence and high support can be considered an insight.

Lift ratio is another measure to judge the importance of a rule. It denotes whether a derived rule is better than the benchmark confidence, that is, the confidence of considering only the consequent. In our previous example, 80% of the entire population buys bread. If we consider the rule with Butter -> Bread, which predicts that the customers will buy bread with 100% confidence, the rule will have a lift of 100/80 = 1.25.

Lift = Rule Confidence/Confidence(Consequent)

When lift ratio is greater than 1, it means that the consequent and antecedent occur more frequently than expected, so we can consider it a good rule.

Association rules are not considered good for problems involving rare events, as the methodology to calculate the rules is highly dependent on the support and confidence measures. These measures are directly proportional to the number of transactions in the data set. So, if the occurrence of certain events is more in the transactions, the better their support, confidence, and lift measures.

Usually, rare events have low occurrence in transactions and are unidentifiable using these measures. These transactions are known as null transactions, as they don't contain the events/items that are examined. So, an optimum plan can be decreasing the support and confidence threshold during the process of building an association rule model and then using null invariant measures to identify these records. Null invariant measures consider only the interaction between the events/items and are not influenced by null transactions. Null invariant measures such as Cosine, Kulc, and Max_Confidence can be used in conjunction with other discussed measures to identify some of the rare, interesting patterns.

In the following sections, I will discuss the association rule mining using Oracle Advanced Analytics technology stack.

Algorithm Settings

Table 4-1 is a list of settings necessary to tune to the Apriori algorithm in Oracle Advanced Analytics.

Table 4-1. *Algorithm Settings*

Parameter	Default Value	Description
Maximum Rule Length	4	Defines the maximum length of a rule. For example, if we specify the maximum rule length as 4, the antecedent will have at most 3 items and the consequent will have 1 item.
Minimum confidence(%)	10	Minimum confidence threshold for rules to be generated.
Minimum support(%)	1	Minimum support threshold for the item sets to be considered as frequent item sets.

Model Settings

To create an association rules model, the attributes in Table 4-2 are required.

Table 4-2. *Model Settings*

Parameter	Description
Transaction ID	Provides the unique identifier that identifies a transaction. It can be a single column or a combination of multiple columns.
Item ID	Name of the column whose data is to be analyzed.
Value	Whether to have existence-based or value-based association rules. Value-based association rules generate rules by taking the occurrence of items and one of their measures such as quantity. Existence-based methods only consider the items.

Association Rules Using SQL and PLSQL

In this section, I will demonstrate the different methods available in Oracle Advanced Analytics to conduct affinity analysis using association rules. The different methods will help you to model a suitable solution based on your use case and the preferred skill set.

> ***STEP 1:*** Create a setting table and insert model settings for association rules. It overrides the default values for the model-setting parameters. This table requires at least one record for odms_item_id_column_name to define the item_id parameter. If other records are not inserting, the model-building API precedes with the default parameters.

The settings table has two attributes:

- setting_name: Name of the setting parameter
- setting_value: Value of the setting parameter

```
set serveroutput on
CREATE TABLE ar_settings_demo (
  setting_name  VARCHAR2(30),
  setting_value VARCHAR2(4000));
/
BEGIN
-- Inserts the default overriden value for minimum support
  INSERT INTO ar_settings_demo VALUES
  (dbms_data_mining.asso_min_support,0.1);
-- Inserts the default overriden value for minimum confidence
  INSERT INTO ar_settings_demo VALUES
  (dbms_data_mining.asso_min_confidence,0.1);
-- Inserts the default overriden value for maximum rule length
  INSERT INTO ar_settings_demo VALUES
  (dbms_data_mining.asso_max_rule_length,3);
-- Inserts the Item ID
  INSERT INTO ar_settings_demo VALUES
```

```
  (dbms_data_mining.odms_item_id_column_name, 'PRODUCT_ID');
  COMMIT;
END;
/
```

STEP 2: Create the mining model.

To create the association rules model, you need to call the CREATE_MODEL Procedure of the DBMS_DATA_MINING package with the parameters shown in Table 4-3.

Table 4-3. *Parameters for Creating an Association Rules Model Using DBMS_DATA_ MINING.CREATE_MODEL*

Parameter	Mandatory	Description
model_name	Yes	To assign a meaningful model name for the model
mining_function	Yes	The data mining function that is to be used; for association rules, you need to call DBMS_DATA_ MINING.ASSOCIATION function
data_table_name	Yes	Name of the data table/view to be used to create the model
case_id_column_name	Yes	The unique record identifier for the data set; the transaction_id is used as a case id column
settings_table_name	Yes	Name of the table that contains the model settings values

The syntax for creating new association rules model using PLSQL API is as follows:

```
BEGIN
  DBMS_DATA_MINING.CREATE_MODEL(
    model_name          => 'DEMO_AR_MODEL',
    mining_function     => DBMS_DATA_MINING.ASSOCIATION,
    data_table_name     => 'DEMO_AR_RULES_DATA',
    case_id_column_name => 'TRANS_ID',
    settings_table_name => 'ar_settings_demo'
    );
END;
/
```

STEP 3: Check if the model is created successfully
(see Figure 4-10).

```
select * from user_mining_models where model_name = 'DEMO_AR_MODEL'
```

125

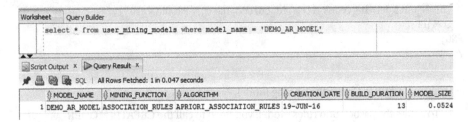

Figure 4-10. User mining model record

> **STEP 4:** Check the model settings that were used to create the
> model (see Figure 4-11).

```
SELECT setting_name, setting_value
  FROM user_mining_model_settings
 WHERE model_name = 'DEMO_AR_MODEL'
ORDER BY setting_name;
```

Script Output × ▷ Query Result ×

SQL | All Rows Fetched: 5 in 0.005 seconds

	SETTING_NAME	SETTING_VALUE
1	ASSO_MAX_RULE_LENGTH	3
2	ASSO_MIN_CONFIDENCE	.1
3	ASSO_MIN_SUPPORT	.1
4	ODMS_ITEM_ID_COLUMN_NAME	PRODUCT_ID
5	PREP_AUTO	OFF

Figure 4-11. Model Settings for association rules model

> **STEP 5:** Result evaluation.

The function GET_ASSOCIATION_RULES of DBMS_DATA_MINING package can
be used to get the generated association rules. The result from this method is of table
datatype (DM_RULES structure); the API has to be invoked as shown in the following
example. The example shows the method to extract rules with the top 10% of item sets
based on the support value of the model (see also Figure 4-12).

```
SELECT a.attribute_subname antecedent,
       c.attribute_subname consequent,
       rule_support support,
       rule_confidence confidence,
       rule_lift lift
  FROM TABLE(DBMS_DATA_MINING.GET_ASSOCIATION_RULES('DEMO_AR_MODEL', 10)) T,
       TABLE(T.consequent) C,
       TABLE(T.antecedent) A
 ORDER BY lift DESC, support DESC, support DESC;
```

	⊕ ANTECEDENT	⊕ CONSEQUENT	⊕ SUPPORT	⊕ CONFIDENCE	⊕ LIFT
1	P04	P02	0.333333333333333333333333333333333333333333	1	1.5
2	P03	P02	0.333333333333333333333333333333333333333333	1	1.5
3	P01	P04	1	1	1
4	P04	P01	1	1	1
5	P02	P04	0.666666666666666666666666666666666666666667	1	1
6	P02	P04	0.666666666666666666666666666666666666666667	1	1
7	P02	P01	0.666666666666666666666666666666666666666667	1	1
8	P04	P01	0.666666666666666666666666666666666666666667	1	1
9	P02	P01	0.666666666666666666666666666666666666666667	1	1
10	P01	P04	0.666666666666666666666666666666666666666667	1	1
11	P03	P04	0.333333333333333333333333333333333333333333	1	1
12	P02	P04	0.333333333333333333333333333333333333333333	1	1
13	P05	P04	0.333333333333333333333333333333333333333333	1	1

Figure 4-12. *Association rules results*

Creating the Association Rules Model Using Oracle R Enterprise

The association rules algorithm is computationally intensive. If you are an R user, you can port your R code easily to the Oracle R Enterprise environment, which will help you leverage the power of in-database computation, making the models run faster. You need to boot up the R client console and follow the steps shown next for creating association rules using Oracle R Enterprise.

STEP 1: Load the ORE and arules library.

```
library(ORE)
library(arules)
```

■ **Note** The arules library is optional. It is useful if you want to analyze the results back in your R client using various native R packages.

STEP 2: Use ore.connect to connect to the database.

```
if (!ore.is.connected()) # Check if client is already connected to R
ore.connect("dmuser", "orcl","localhost", "sibanjan123", all=TRUE)
```

STEP 3: Read the CSV (comma-separated values) file using the read.csv function. This converts the data into a data frame in R.

```
assoc_df <- read.csv("C:/Users/Admin/Dropbox/analytics_book/chapter-5/
example_data.csv")
```

STEP 4: Later this data will be pushed to the Oracle database for embedded R execution. The attributes' names should be in uppercase to be recognized by the model building APIs. Rename the attributes of the data frame to uppercase using native R's names() function.

```
names(assoc_df)<-c("TRANS_ID","PRODUCT_ID")  # Convert to uppercase so that
API recognizes it as a table column name
```

STEP 5: Use ore.push function to translate the R object to corresponding ore.frame object; ore.frame maps native R objects to Oracle R specific database objects.

```
ore.assoc_df <-ore.push(assoc_df)
```

STEP 6: Invoke ore.odmAssocRules with the model-setting parameters to create the association rules model.

```
ar.mod <-ore.odmAssocRules(~., ore.assoc_df, case.id.column= "TRANS_
ID",item.id.column= "PRODUCT_ID", min.support= 0.5, min.confidence= 0.7,max.
rule.length = 3)
```

STEP 7: Use the itemsets function to read and view the generated item sets (see also Figure 4-13).

```
itemsets<-itemsets(ar.mod)
itemsets
```

```
> itemsets
   ITEMSET_ID NUMBER_OF_ITEMS ITEMS    SUPPORT
1           1               1     P01 1.0000000
2           3               1     P04 1.0000000
3           5               2     P01 1.0000000
4           5               2     P04 1.0000000
5           2               1     P02 0.6666667
6           4               2     P01 0.6666667
7           4               2     P02 0.6666667
8           6               2     P02 0.6666667
9           6               2     P04 0.6666667
10          7               3     P01 0.6666667
11          7               3     P02 0.6666667
12          7               3     P04 0.6666667
```

Figure 4-13. Itemsets

STEP 8: Use the rules function to view the generated association rules (see Figure 4-14).

```
rules <-rules(ar.mod)
rules
```

```
> rules
  RULE_ID NUMBER_OF_ITEMS LHS RHS   SUPPORT CONFIDENCE LIFT
1       2               1 P01 P04 1.0000000          1    1
2       3               1 P04 P01 1.0000000          1    1
3       4               1 P02 P04 0.6666667          1    1
4       1               1 P02 P01 0.6666667          1    1
5       5               2 P02 P04 0.6666667          1    1
6       5               2 P01 P04 0.6666667          1    1
7       6               2 P04 P01 0.6666667          1    1
8       6               2 P02 P01 0.6666667          1    1
```

Figure 4-14. *Association rules*

> STEP 9: Sometimes there is a need to select observations
> specific to some attributes. We can use the subset function
> to pull out those specific rules. The following code example
> displays only the rules that have P04 as a consequent (see also
> Figure 4-15).

```
#Subset rules with item P04
sub.rules<-subset(rules, min.confidence=0.7, rhs=list("P04"))
sub.rules
```

```
> sub.rules
  RULE_ID NUMBER_OF_ITEMS LHS RHS   SUPPORT CONFIDENCE LIFT
1       2               1 P01 P04 1.0000000          1    1
2       5               2 P01 P04 0.6666667          1    1
3       5               2 P02 P04 0.6666667          1    1
4       4               1 P02 P04 0.6666667          1    1
```

Figure 4-15. *Rules subset*

> STEP 10: Native R users enjoy analyzing the results using the
> many open source functions that R offers. You can pull out
> the results from the database to an R frame using the ore.pull
> function. The following code pulls out the results from the
> database and inspects the rules using the inspect function
> (see also Figure 4-16).

```
# Convert the rules to the rules object of arulespackage
rules.arules<-ore.pull(rules)
inspect(rules.arules)
```

```
> inspect(rules.arules)
   lhs         rhs       support confidence lift
1 {P01} => {P04} 1.0000000          1    1
2 {P04} => {P01} 1.0000000          1    1
3 {P02} => {P04} 0.6666667          1    1
4 {P02} => {P01} 0.6666667          1    1
5 {P01,
   P02} => {P04} 0.6666667          1    1
6 {P02,
   P04} => {P01} 0.6666667          1    1
```

Figure 4-16. *Inspect rules*

Creating the Association Model Using SQL Developer

Create a data mining project and a workflow. Refer to the steps discussed in Chapter 3 to create a new project and workflow. Once a new workflow is created, follow the steps shown here to create a cluster model in SQL Developer.

> ***STEP 1:*** Create a data source for table DEMO_AR_RULES_ DATA (see Figure 4-17). Refer to Chapter 3 for the details on creating a data source.

Figure 4-17. *Create a Data Source node*

> ***STEP 2:*** To build an association rule model, select and drag the Assoc Build from the Model section in Components editor to the worksheet. Connect the data source tab and the Assoc Build Model as shown in Figure 4-18.

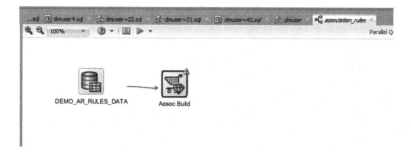

Figure 4-18. *Connect Assoc Build node to the DEMO_AR_RULES_DATA node*

> **STEP 3:** Edit the settings of the Assoc Build node. Right-click and select Edit from the list (see Figure 4-19).

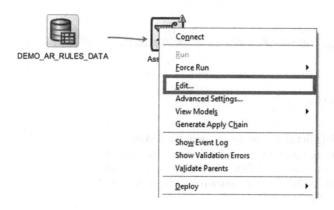

Figure 4-19. *Edit Assoc Build node*

> **STEP 4:** Enter the required details: Transaction ID and Item ID of the data set. The value field is optional and defaults with Existence (see Figure 4-20).

Figure 4-20. *Association Build parameters*

> **STEP 5:** Model settings have the parameters to enter
> maximum rule length, minimum support, and confidence
> thresholds. To edit the model settings, click on the pencil icon
> as shown in Figure 4-21.

Figure 4-21. *Edit Model Settings*

> **STEP 6:** The default values are already populated in the
> algorithm settings. Enter the values for the model settings of
> your choice to override the defaults. In this example, we enter
> 4 for maximum rule length to generate rules with maximum 3
> items in a rule (Figure 4-22).

Figure 4-22. *Algorithm Settings*

> **STEP 7:** Execute the workflow and wait until it completes
> successfully (Figure 4-23).

DEMO_ASSOC_RULES_TAB   Assoc Build

Figure 4-23. *Final workflow*

STEP 8: Result Evaluation.

To view the results, right-click on the Assoc Build node and select the model from the view model option.

Table 4-4. *Result Fields*

Column	Description
ID	Record Identifier
Antecedent	Rule antecedent items
Consequent	Rule Consequent items
Lift	Lift of the rule
Confidence(%)	Rule Confidence
Support(%)	Support Confidence
Item Count	Number of antecedent items included in the rule
Antecedent Support(%)	Support of the rule antecedent
Consequent Support(%)	Support of the rule consequent

Once the model result appears on the screen, the rules along with the information listed in Table 4-4 appear on the screen (see Figure 4-24).

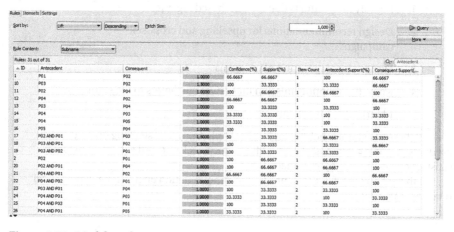

Figure 4-24. *Model results*

STEP 9: Filtering rules with specified criteria.

You can use the More tab to filter rules based on filter criteria such as Minimum Lift, Minimum Support(%), Minimum Confidence(%), Maximum Items In Rule, and Minimum Items In Rule (Figure 4-25).

Figure 4-25. *Rules filter options*

Case Study—Market Basket Analysis

"I don't see smiles on the faces of people at Whole Foods."

—Warren Buffett

Market basket analysis is often used by retailers to know what their customers buy, and they use this information to make customers happy. Every shopper's basket has a lot of information and a story to tell. Whether the customer has a weird choice of products or the customer is in some sort of shopping mission are some of the areas that can only be known from the products they stock in their shopping baskets. Use of association rule analysis is a core module for market basket analysis to discover various kinds of insights out of the data. A retailer can use this information to

1. Identify the products that are purchased together

2. Design recommendations for upselling and cross-sell campaigns

3. Design store layout and planograms

4. Know the shopping mission that drives customers to the store

5. Identify the driver items

All these insights are helpful for inventory planning and stocking up on the right products that are in demand. This avoids product stock out, and achieves customer satisfaction and greater operational efficiency.

In this section, I will create a product recommendation engine for a hypothetical retail giant Walstar Inc.

Business Understanding

Walstar is a multinational retail corporation that runs chains of large department stores. Over the last decade, the company had lost customers to new entrants in the market. The company is well known for its low pricing and high-value products. Recently, Walstar's profits have slid by 7%, and its same-store sales are falling. To maximize its profit, Walstar's business analysts believe it is essential to maximize the sales revenue and minimize all associated costs. Some of the methods that the team has identified to increase sales revenue include

- Retaining the existing customers

- Growing the share of the wallet

- Improving the impact of promotional campaigns and better meeting customer expectations

All these goals are better aligned when the customers are lured to buy more items from the store. You, being a data scientist, suggested building a product recommendation engine for Walstar Inc. Nowadays, recommendation engines are quite common for online stores. However, for physical retail stores, the point where personalized recommendations can be given is at the POS (point-of-sale) counters. POS counters are where a retail transaction is completed. Therefore, the designed recommendation engine has to be associated with some discount to make the customer remember to return to the store and make the purchase. Providing relevant recommendations at the customer touch points is a key driver in retaining existing customers.

The recommendation engine to be developed for Walstar Inc. attempts to maximize sales, integrating product information with customer data to offer relevant and personalized recommendations.

The developed product recommender module should also be agile at decision-making, as it is to be installed at the POS counters during the customer checkout process. Based on the current purchase by the customer, the system would recommend a list of products, which are associated with discounts. There would also be a discount's expiration date to increase the probability of a customer returning for the purchase. This also helps to have an estimated date of purchase, which can be used for planning the inventory. The salient features of the product recommender module would be the ability to seamlessly integrate with the order a management module that runs on the Oracle database.

Data Understanding

The team at Walstar Inc. liked the idea of a product recommendation engine. You are provided access to their Order Management database. Table 4-5 shows a brief of the metadata of the useful tables in their sales database:

Table 4-5. Metadata for Sales history

Table	Description
DEMO_PURCHASE_TRANS	Contains the history of purchase transactions
DEMO_PURCHASE_PRODUCTS	Has product details information
DEMO_STORES	Has information on Walstar's stores

Data Preparation

The following is a list of attributes that are required to create a product recommendation engine for Walstar:

- Transaction Id: To uniquely identify each purchase transaction

- Item: The sold products

- Purchase Quantity: The quantity of products sold

The following attributes are not required for market basket analysis but for building the personalized product recommendation engine:

- Customer Id: Unique customer identifier

- Store Id: To identify the store where the purchase was made

Based on this knowledge, you create a database view DEMO_ASSOC_PUR_V that provides you a result set with the previously mentioned attributes. (see Figure 4-26)

```
CREATE OR REPLACE VIEW DEMO_ASSOC_PUR_V AS
select dpt.transaction_id,dpt.customer_id,dpt.store_id,dpp.item,dpt.purchase_qty
from  DEMO_PURCHASE_TRANS dpt,
      DEMO_PURCHASE_PRODUCTS dpp
  where dpt.item_id=dpp.item_id
```

Figure 4-26. View definition for DEMO_ASSOC_PUR_V

Data Modeling

A two-step process as shown in Figure 4-27 is designed to solve the problem of building a robust recommendation system. First, a market basket analysis is carried out to find the product association rules and this is stored in Walstar's database system. Next, the rules along with other sales data are used at the POS in real time to recommend products to the customers.

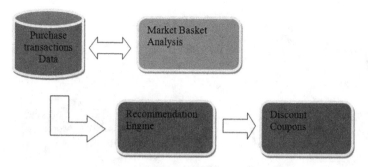

Figure 4-27. *High-level architecture of product recommender engine*

This two-step process would help to target the customers better by a lift in the purchase probability, which would eventually help to reduce the costs associated with the promotional discounts and marketing efforts.

■ **Note** In this chapter, I will go through the market basket analysis part of the recommendation engine. I will discuss on the second part when we walk through the recommendation engines in Chapter 7.

The application demonstration is coded using PLSQL and Oracle R Enterprise to showcase the capabilities of PLSQL with R integration. This can also be prepared using SQL Developer GUI.

High-Level Technical Overview

A PLSQL package DEMO_ASSOC_MODELS_PUB is developed to prepare the market basket analysis module of the product recommendation engine. As shown in Figure 4-28, it has the components to build models and insert the results into a database table.

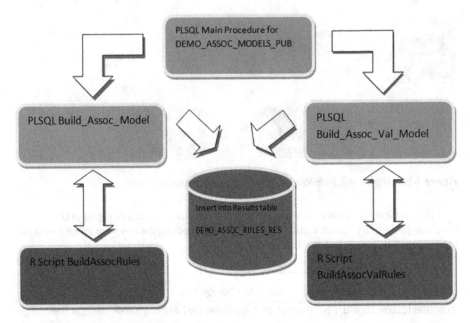

Figure 4-28. *High-level technical architecture*

The main procedure of DEMO_ASSOC_MODELS_PUB package drives the application. It accepts one parameter (p_model_type), which is required to conditionally decide whether to execute an existence-based or a value-based association rule model.

```
Procedure MAIN(p_model_type In Varchar2)
IS
    l_method_name    VARCHAR2(30):= 'Build_Assoc_Model';
        l_data_view      VARCHAR2(200);
        l_return_status  VARCHAR2(10);
        l_return_msg     VARCHAR2(4000);
BEGIN

        LOG_MESSAGE('Start of '||G_PACKAGE_NAME||'.'||l_method_name);
-- Assign the view name to the l_data_view variable
        l_data_view :='DEMO_ASSOC_PUR_V';

-- Condition to check model_type: Value/Existence
        IF p_model_type='VALUE' then

-- Invoke the Build_Assoc_Val_Model for value based model
            Build_Assoc_Val_Model(p_data_view => l_data_view,
                                  x_return_status => l_return_status,
                                  x_return_msg    => l_return_msg);
        ELSE
```

```
-- Invoke the Build_Assoc_Model for existence based model
              Build_Assoc_Model(p_data_view => l_data_view,
                                 x_return_status => l_return_status,
                                 x_return_msg    => l_return_msg);
        END IF;
    -- Check the return status
    IF l_return_status='S' then
              LOG_MESSAGE('Rules were generated successfully and stored in
              database');
        ELSE
              LOG_MESSAGE('Unsuccessful. Please check log or debug the
              code. The error message was '||l_return_msg);
        END IF;

              LOG_MESSAGE('End of '||G_PACKAGE_NAME||'.'||l_method_name);

END MAIN;
```

Procedure Build_Assoc_Model has the functions to execute the existence-based association model, interface the PLSQL package to Oracle R Enterprise environment, and insert the output to the results table in the Oracle Database. It accepts parameters described in Table 4-6.

Table 4-6. *Parameters for Build_Assoc_Model Procedure*

Parameter Name	Description
p_data_view	The name of a database view, i.e., DEMO_ASSOC_PUR_V
x_return_status	The return status that notifies whether there were any exceptions(E) in the execution or if it completed successfully(S)
x_return_msg	The error message in case of any exceptions

```
Procedure Build_Assoc_Model(
      p_data_view            In Varchar2,
      x_return_status        Out Nocopy Varchar2,
      x_return_msg           OUT NOCOPY VARCHAR2
   ) IS
        l_method_name        VARCHAR2(30):= 'Build_Assoc_Model';
        l_output_table       VARCHAR(30) := 'DEMO_ASSOC_RULES_RES';
        l_insert_sql         VARCHAR2(32767);
        l_last_update_date   DATE    := SYSDATE;
        l_last_updated_by    NUMBER  := -1;
        l_creation_date      DATE    := SYSDATE;
        l_created_by         NUMBER  := -1;
l_mining_function    VARCHAR2(30) := 'BuildAssocRules';

    BEGIN
```

```
      LOG_MESSAGE('START OF '||G_PACKAGE_NAME||'.'||l_method_name);

-- Invokes ORE function BuildAssocRules and insert the results into table
DEMO_ASSOC_RULES_RES

l_insert_sql := 'INSERT INTO '||l_output_table||'(RESULT_ID,MODEL_TYPE,LAST_
UPDATE_DATE,LAST_UPDATED_BY,CREATION_DATE,CREATED_BY,RULES,SUPPORT,CON
FIDENCE,LIFT) '||' SELECT DEMO_ASSOC_RULES_S.NEXTVAL'||',''EXISTENCE''
,'''||l_last_update_date||''','||l_last_updated_by||',''''||l_creation_
date||''','||l_created_by||',rqtab.* FROM table(rqTableEval(cursor(SELECT
* FROM '||p_data_view||'),cursor(SELECT 1 as "ore.connect" FROM
dual),'' select cast(''''a'''' as VARCHAR2(4000)) "RULES",1 "SUPPORT", 1
"CONFIDENCE", 1 "LIFT" from dual'',''''||l_mining_function||''''')) rqtab';

LOG_MESSAGE(G_PACKAGE_NAME||'.'||l_method_name||' Dynamic insert into '||l_
output_table||' Query: '||l_insert_sql);

-- Executes the l_insert_sql statemetn
EXECUTE IMMEDIATE l_insert_sql;

COMMIT;

x_return_status := 'S';

LOG_MESSAGE('END OF '||G_PACKAGE_NAME||'.'||l_method_name);

EXCEPTION WHEN OTHERS THEN
x_return_status := 'E';
x_return_msg := G_PACKAGE_NAME||'.'||l_method_name||': '||SQLERRM;
END Build_Assoc_Model;
```

Procedure Build_Assoc_Model uses rqTableEval to invoke the created ORE function BuildAssocRules. rqTableEval is a PLSQL interface function that invokes R scripts with a database table as an input. Its syntax is as follows:

```
SELECT *
  FROM table(rqTableEval(
<Input_Cursor>,
<Parameter_Cursor>,
<Output_Query>,
<Script_Name>));
```

The parameters for the rqTableEval interface function are described in Table 4-7.

Table 4-7. *Parameters for rqTableEval Interface Function*

Parameter Name	Description
Input_Cursor	Cursor that specifies the data to be used for building the models
Parameter_Cursor	List of values for the arguments specified in the ORE function
Output_Query	One of the following values: An SQL select statement to specify the column name and datatype of the returned results NULL to return a serialized data object XML to return the results in an XML file format
Script_Name	The Name of the ORE script

Procedure Build_Assoc_Val_Model has the same logic as Build_Assoc_Model procedure. However, this procedure is for creating value-based association rules.

```
Procedure Build_Assoc_Val_Model(
    p_data_view              In Varchar2,
    x_return_status          Out Nocopy Varchar2,
    x_return_msg             OUT NOCOPY VARCHAR2
)
IS
        l_method_name        VARCHAR2(30):= 'Build_Assoc_Val_Model';
        l_output_table       VARCHAR(30) := 'DEMO_ASSOC_RULES_RES';
        l_insert_sql         VARCHAR2(32767);
        l_last_update_date   DATE    := SYSDATE;
        l_last_updated_by    NUMBER  := -1;
        l_creation_date      DATE    := SYSDATE;
        l_created_by         NUMBER  := -1;
l_mining_function    VARCHAR2(30) := 'BuildAssocValRules';

    BEGIN

        LOG_MESSAGE('START OF '||G_PACKAGE_NAME||'.'||l_method_name);

-- Invokes ORE function BuildAssocValRules and insert the results into table
DEMO_ASSOC_RULES_RES

l_insert_sql := 'INSERT INTO '||l_output_table||'(RESULT_ID,MODEL_TYPE,LAST_
UPDATE_DATE,LAST_UPDATED_BY,CREATION_DATE,CREATED_BY,RULES,SUPPORT,CONFIDEN
CE,LIFT) '||' SELECT DEMO_ASSOC_RULES_S.NEXTVAL||',''VALUE'','''||l_last_
update_date||''','||l_last_updated_by||','''||l_creation_date||''','''||l_
created_by||',rqtab.* FROM table(rqTableEval(cursor(SELECT * FROM
'||p_data_view||'),cursor(SELECT 1 as "ore.connect" FROM dual),'' select
cast(''''a'''' as VARCHAR2(4000)) "RULES",1 "SUPPORT", 1 "CONFIDENCE", 1
"LIFT" from dual'','''||l_mining_function||''')) rqtab';
```

```
LOG_MESSAGE(G_PACKAGE_NAME||'.'||l_method_name||' Dynamic insert into '||l_
output_table||' Query: '||l_insert_sql);

EXECUTE IMMEDIATE l_insert_sql;

COMMIT;

x_return_status := 'S';

LOG_MESSAGE('END OF '||G_PACKAGE_NAME||'.'||l_method_name);

EXCEPTION WHEN OTHERS THEN
x_return_status := 'E';
x_return_msg := G_PACKAGE_NAME||'.'||l_method_name||': '||SQLERRM;

END Build_Assoc_Val_Model;
```

BuildAssocValRules is the ORE procedure that holds the R script for building the value-based association rules model. The attribute PURCHASE_QTY, which is assigned to the item.value.column parameter, is used to create the value-based model. The model will create rules with insights such as a customer who buys six breads with two butters also buys five avocado having 10% support and 50% confidence. Here, six, two, and five are the purchase quantity.

```
begin
sys.rqScriptDrop('BuildAssocValRules');
sys.rqScriptCreate('BuildAssocValRules',
'function(dat, min.support=0.01, min.conf = 0.1, max.run.len=5)
{
library(ORE)
library(arules)
dataODF <- as.ore.frame(dat)
assoc.mod <- ore.odmAssocRules(~., dataODF,case.id.column= "TRANSACTION_
ID",item.id.column= "ITEM", item.value.column="PURCHASE_QTY",min.support=
min.support,min.confidence= min.conf ,max.rule.length = max.run.len)
rules <-rules(assoc.mod)
rules <-ore.pull(rules)
as(rules, "data.frame")
}');
END;
```

BuildAssocRules is the ORE procedure that is an R script for the building association rules model. In this procedure, the rules will be created based on the products brought. For example, rules such as the customer who brought bread also brought butter with 20% support and 70% confidence would be formed.

```
begin
-- sys.rqScriptDrop drops the existing ORE script with the same name. This
script has to be commented out on the first run
sys.rqScriptDrop('BuildAssocRules');
-- sys. rqScriptCreate to create a new ORE script
sys.rqScriptCreate('BuildAssocRules',
'function(dat, min.support=0.01, min.conf = 0.1, max.run.len=5)
{
library(ORE)
library(arules)
dataODF <- as.ore.frame(dat)
#ore.odmAssocRules is the ORE equivalent function for ODM association rules.
The model and
#algorithims settings are similar to that described in above sections
assoc.mod <- ore.odmAssocRules(~., dataODF,case.id.column= "TRANSACTION_
ID",item.id.column= "ITEM", min.support= min.support,min.confidence= min.
conf ,max.rule.length = max.run.len)
rules <-rules(assoc.mod)
rules <-ore.pull(rules)
#Converts the rules to a data frame to be recognized by rqTableEval as
individual columns and
#inserted into the results table
as(rules, "data.frame")
}');
END;
```

Execution

Run the MAIN procedure of the DEMO_ASSOC_MODELS_PUB with VALUE as the model
parameter to run the value-based association rule model as shown in Figure 4-29.

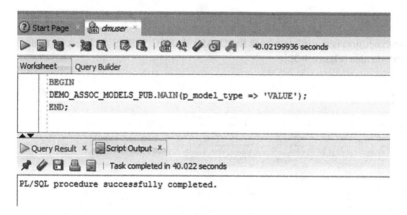

Figure 4-29. *Execute the DEMO_ASSOC_MODELS package to create association rules*

Validate association rules from the DEMO_ASSOC_RULES_RES results table. The MODEL_TYPE column (see Figure 4-30) shows the newly created value-based rules. Also the results display the rules in an {Item = Quantity} format such as {BBQ sauce=1}.

	RESULT_ID	MODEL_TYPE	LAST_UPDATE_DATE	LAST_UPDATED_BY	CREATION_DATE	CREATED_BY	RULES	SUPPORT	CONFIDENCE	LIFT
1	2041	VALUE	19-JUN-16	-1	19-JUN-16	-1	{BBQ sauce=1,BBQ sauce=5,BBQ sauce=6} => {BBQ sauce=2}	0.0122399020807834	0.625	
2	2042	VALUE	19-JUN-16	-1	19-JUN-16	-1	{BBQ sauce=1,BBQ sauce=5,Bread=5} => {BBQ sauce=4}	0.0122399020807834	0.625	15.018982:
3	2043	VALUE	19-JUN-16	-1	19-JUN-16	-1	{BBQ sauce=1,BBQ sauce=5,Bread=5} => {BBQ sauce=2}	0.0122399020807834	0.625	
4	2044	VALUE	19-JUN-16	-1	19-JUN-16	-1	{BBQ sauce=1,BBQ sauce=5,BBQ sauce=6} => {Bread=6}	0.0122399020807834	0.625	12
5	2045	VALUE	19-JUN-16	-1	19-JUN-16	-1	{BBQ sauce=4,Bread=6} => {BBQ sauce=5}	0.0122399020807834	0.625	10.012254:
6	2046	VALUE	19-JUN-16	-1	19-JUN-16	-1	{BBQ sauce=2,Bread=6} => {BBQ sauce=1}	0.0122399020807834	0.625	11.605113:
7	2047	VALUE	19-JUN-16	-1	19-JUN-16	-1	{BBQ sauce=1,Bread=6} => {BBQ sauce=2}	0.0122399020807834	0.625	
8	2048	VALUE	19-JUN-16	-1	19-JUN-16	-1	{Bread=6,Bread=6} => {BBQ sauce=6}	0.0122399020807834	0.625	12.157738(
9	2049	VALUE	19-JUN-16	-1	19-JUN-16	-1	{Bread=6,Bread=6} => {BBQ sauce=5}	0.0122399020807834	0.625	10.012254:
10	2050	VALUE	19-JUN-16	-1	19-JUN-16	-1	{BBQ sauce=2,Bread=6} => {Bread=5}	0.0122399020807834	0.625	10.636020(
11	2051	VALUE	19-JUN-16	-1	19-JUN-16	-1	{Bread=5,Bread=6} => {BBQ sauce=2}	0.0122399020807834	0.625	
12	2052	VALUE	19-JUN-16	-1	19-JUN-16	-1	{BBQ sauce=5,Bread=4} => {BBQ sauce=3}	0.0122399020807834	0.625	15.018982:
13	2053	VALUE	19-JUN-16	-1	19-JUN-16	-1	{BBQ sauce=4,Bread=6} => {Bread=4}	0.0122399020807834	0.625	12
14	2054	VALUE	19-JUN-16	-1	19-JUN-16	-1	{BBQ sauce=2,Bread=6} => {Bread=4}	0.0122399020807834	0.625	12
15	2055	VALUE	19-JUN-16	-1	19-JUN-16	-1	{BBQ sauce=6,Bread=1} => {BBQ sauce=4}	0.0122399020807834	0.625	15.018982:
16	2056	VALUE	19-JUN-16	-1	19-JUN-16	-1	{BBQ sauce=1,BBQ sauce=6} => {Bread=4}	0.0159118727050184	0.619047619047619	12.644047(
17	2057	VALUE	19-JUN-16	-1	19-JUN-16	-1	{BBQ sauce=5,BBQ sauce=6} => {BBQ sauce=4}	0.0159118727050184	0.619047619047619	14.875356

Figure 4-30. *Results from the DEMO_ASSOC_RULES_RES table*

Similarly, execute the procedure with p_model_type parameter as EXISTENCE (Figure 4-31).

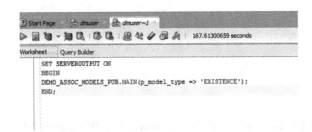

```
SET SERVEROUTPUT ON
BEGIN
DEMO_ASSOC_MODELS_PUB.MAIN(p_model_type => 'EXISTENCE');
END;
```

Figure 4-31. *Execute DEMO_ASSOC_MODELS_PUB for existence type of association rules models*

Once the execution is complete, check the DEMO_ASSOC_RULES_RES results table for the generated rules (Figure 4-32).

	RESULT_ID	MODEL_TYPE	LAST_UPDATE_DATE	LAST_UPDATED_BY	CREATION_DATE	CREATED_BY	RULES	SUPPORT	CONFIDENCE	LIFT
388	1660	EXISTENCE	19-JUN-16	-1	19-JUN-16	-1	{BBQ sauce,Mushrooms} => {Bread}	0.0428396572927417	1	5.7535211:
389	1661	EXISTENCE	19-JUN-16	-1	19-JUN-16	-1	{Beets,Mushrooms} => {Bread}	0.0416156670746634	1	5.7535211:
390	1662	EXISTENCE	19-JUN-16	-1	19-JUN-16	-1	{Beets,Bread,Mushrooms} => {BBQ sauce}	0.0416156670746634	1	
391	1663	EXISTENCE	19-JUN-16	-1	19-JUN-16	-1	{BBQ sauce,Beets,Mushrooms} => {Bread}	0.0416156670746634	1	5.7535211:
392	1664	EXISTENCE	19-JUN-16	-1	19-JUN-18	-1	{BBQ sauce,Cucumbers} => {Bread}	0.0416156670746634	1	5.7535211:
393	1665	EXISTENCE	19-JUN-16	-1	19-JUN-16	-1	{BBQ sauce,Peppers} => {Bread}	0.0416156670746634	1	5.7535211:
394	1666	EXISTENCE	19-JUN-16	-1	19-JUN-16	-1	{BBQ sauce,Carrots} => {Bread}	0.0416156670746634	1	5.7535211:
395	1667	EXISTENCE	19-JUN-16	-1	19-JUN-16	-1	{Beets,Mushrooms} => {BBQ sauce}	0.0416156670746634	1	
396	1668	EXISTENCE	19-JUN-16	-1	19-JUN-16	-1	{Beets,Carrots} => {Bread}	0.0403914768665851	1	5.7535211:
397	1669	EXISTENCE	19-JUN-16	-1	19-JUN-16	-1	{Beets,Bread,Cucumbers} => {BBQ sauce}	0.0403914768665851	1	
398	1670	EXISTENCE	19-JUN-16	-1	19-JUN-16	-1	{BBQ sauce,Beets,Cucumbers} => {Bread}	0.0403914768665851	1	5.7535211:
399	1671	EXISTENCE	19-JUN-16	-1	19-JUN-16	-1	{Beets,Bread,Peppers} => {BBQ sauce}	0.0403914768665851	1	
400	1672	EXISTENCE	19-JUN-16	-1	19-JUN-16	-1	{BBQ sauce,Beets,Peppers} => {Bread}	0.0403914768665851	1	5.7535211:
401	1673	EXISTENCE	19-JUN-16	-1	19-JUN-16	-1	{Beets,Bread,Carrots} => {BBQ sauce}	0.0403914768665851	1	
402	1674	EXISTENCE	19-JUN-16	-1	19-JUN-16	-1	{BBQ sauce,Beets,Carrots} => {Bread}	0.0403914768665851	1	5.7535211:
403	1675	EXISTENCE	19-JUN-16	-1	19-JUN-16	-1	{Bread,Hot sauce} => {BBQ sauce}	0.0403914768665851	1	
404	1676	EXISTENCE	19-JUN-16	-1	19-JUN-16	-1	{BBQ sauce,Hot sauce} => {Bread}	0.0403914768665851	1	5.7535211:

Figure 4-32. *Results from DEMO_ASSOC_RULES_RES table*

Congratulations!! You have successfully created various types of association rules models and successfully stored the results back to the database tables. We can tie these rules to other business operations database tables to generate various kinds of insights.

Summary

The purpose of this chapter was to introduce you to association rules mining, which is another unsupervised data science technique that Oracle Advanced Analytics offers. Association rules are widely used in various industries to discover the relationships between different attributes. You interactively studied the Apriori algorithm in detail and the various ways through which this algorithm can be executed from different Oracle-based platforms. You also learned about the value-based and existence-based association rules and how they can be modeled using Oracle Advanced Analytics. I discussed market basket analysis, which is one of the widely used applications of association rules in the retail industry and successfully automated the modeling process from data extraction to the storage of results in the database. In Chapter 7, I will be using these rules to prepare a product recommendation engine.

CHAPTER 5

■ ■ ■

Regression Analysis

Knowing the future is something everyone wishes for. People are obsessed with knowing the future, believing that if they know tomorrow, they will adjust their activities today. It is impossible to get data from the future, but there is a way to know the future from reviewing the past. Investors always review past pricing history to influence their future investment decisions. I recently bought a house. Before buying, I wanted to know if it was a good decision to buy. Can I wait for prices to fall? Will the price appreciate in the future? I did a detailed survey of the area and the desirable features such as size of the bedrooms and the price per square foot for individual houses in various areas. This helped me to understand the factors that influence the price in different areas and also what would it be like in the years ahead. Similarly, today when I am writing this chapter, I don't know whether it will be published or not.

Examining the past results and the current commitment from the whole team, I think that it has a higher chance to get published. This is a prediction. Prediction is something that we use to foresee the future from similar past behavior, but there is a chance attached to it. It is not 100% accurate but provides one with an estimate. In the world of business, it helps in proactive planning. In day-to-day operations, we hear questions such as How many customers will we have next quarter? What will be the sales projection for next year? With a few attributes and data points, we can connect the dots by hand and provide a prediction. In business and the real world, we have thousands of attributes and millions of data, so how do we predict? How do we derive a relationship? Here, the opportunity lies in applying supervised machine learning algorithms. It helps us in two ways:

- The algorithms study the past data and can be used to predict the future.

- The algorithms help in determining the relationship between the variable of interest (target/dependent variable) and the factors that influence it (predictors/independent variables)

However, not all supervised algorithms can be applied to a data set. The correct application and the right technique on a polished data set yields expected results.

In this chapter, I will discuss the following topics:

- Understanding relationships

- Introduction to regression analysis

- OLS (ordinary least squares) regression fundamentals

© Sibanjan Das 2016
S. Das, *Data Science Using Oracle Data Miner and Oracle R Enterprise*,
DOI 10.1007/978-1-4842-2614-8_5

- OLS regression using Oracle Advanced Analytics

- GLM and ridge regression overview

- GLM regression using Oracle SQL and PLSQL APIs

- GLM regression using Oracle R Enterprise

- GLM regression using Oracle SQL Developer GUI

- Case Study: Sales Forecasting

Understanding Relationships

Suppose we are a detergent manufacturing company, and we have two variables—the number of units of detergent produced (X) and their cost of production (Y). We want to understand how they are related and whether we can predict cost from the number of produced detergent units. Figure 5-1 lists the data for our analysis.

Production(X)	Cost(Y)
48	6.1
54	6.6
57	6.8
62	7
68	7.1

Figure 5-1. Sample detergent production data

To understand the relationship between two attributes, the first step is to figure out whether they are related. The easiest way is to represent the data visually using graphs. The plot in Figure 5-2 with production in the x axis and cost in y axis illustrates a relationship between these two variables of interest. However, it is just helpful for visualization. It doesn't help to quantify the degree or strength of a relationship between the two variables.

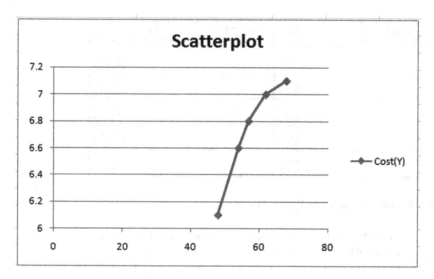

Figure 5-2. *Scatterplot of production(x) and cost(Y)*

The strength and direction of a relationship can be gauged by covariance and correlation measures. Covariance measures the degree to which two variables are related. The statistical formula for covariance is as follows:

$$\text{Covariance}(Y,X) = \frac{\sum_{i=1}^{n}(Xi - \bar{X})(Yi - \bar{Y})}{n-1}$$

If Covariance(X,Y) is greater than zero, the variables are positively related; and if it is less than zero, they are negatively related.

We can calculate covariance for our example data set by following these steps:

STEP 1: Calculate the mean of X and Y (Figure 5-3).

x̄	57.8
ȳ	6.72

Figure 5-3. *Mean(X) and mean(Y)*

STEP 2: Calculate the deviations of actual values from the mean (Figure 5-4).

(Yi-ȳ)*(Xi - X̄)	(Xi - X̄)*(Xi - X̄)
6.076	96.04
0.456	14.44
-0.064	0.64
1.176	17.64
3.876	104.04

Figure 5-4. *Deviation of X and Y from their mean*

STEP 3: Calculate the covariance measure using the equation in Figure 5-4 (see Figure 5-5).

Covariance	2.88

Figure 5-5. *Covariance*

As the covariance measure is greater than zero, we can conclude that the number of units of detergent produced and their cost of production are positively related.

■ **Note** In SQL, the COVAR_POP function can be used to find covariance between two attributes.

However, covariance doesn't provide us with the strength of the relationship, as it is affected by the unit of measurement. For example, the production cost is in dollars and the number of detergent produced is just a count. To measure the strength, they need to be brought to the same scale, that is, standardized. Correlation is another statistical measure that is a standardized version of covariance. It measures both the strength and direction of the relationship. The correlation between the two variables X and Y is given by the following equation:

$$Corrrelation(Y,X) = \frac{Covariance(Y,X)}{Std.Deviation(X) \times std.Deviation(Y)}$$

Correlation always varies between –1 and +1. The closer the value of the correlation coefficient is to 1 or –1, the stronger the relationship. The sign of the correlation result indicates the direction of the relationship. The variables are positively correlated when the correlation value is greater than zero; and if it is less than zero, they are negatively correlated.

We extend the relationship analysis of our example data set by calculating the correlation measure.

> **STEP 4:** Calculate the correlation measure using the previous equation (see Figure 5-6 for the result).

Correlation	0.049484536

Figure 5-6. *Correlation*

A correlation of 0.049 indicates that the relationship between units of detergent produced and their cost of production is positive. However, the strength of their relationship is low.

■ **Note** In SQL, the CORR function can be used to find the correlation between two attributes.

This correlation helped us to quantify the degree to which two variables are related. It didn't provide us with a solution to help us predict the cost from the production units. The degree of relationship measured by correlation might be a chance. We cannot judge the cause and effect relationship between X and Y using correlation. To measure the relationship between the variables and predict cost from production units, we can use regression techniques.

Regression Analysis

Regression analysis is a dominated and highly used statistical method for predictions. This branch of supervised technique works on data sets having continuous target variables. It is termed a horsepower technique in data science and is the first algorithm that data scientists try when they have a target label with continuous numeric values. It is an approach to express the relationship between a target variable (dependent) and one or more independent variables (predictors). The derived relationship helps to predict the unknown values of the target variable. Mathematically, if X is an independent variable and Y is a dependent variable, the relationship is expressed as

$$Y = a + bX + \varepsilon$$

where a is the Intercept of the best-fit regression line, b is slope of the line, and ε is the deviation of the actual and predicted values.

This equation helps us to understand the relationship between X and Y. For example, we can infer

- How much X affects Y

- The change in Y for a given value of X

Also, we can calculate the value of Y for given values of a, b, X, and ε. This is known as predicting an unknown Y from a batch of existing data for X.

We just discussed the equation for simple linear regression, as it involves only one predictor (X) and one target (Y). When there are multiple predictors involved to predict a target, it is known as multiple linear regression. The term "linear" suggests there is a basic assumption that the underlying data exhibits a linear relationship.

If we extend our previous scatter plot and add a trend line to it, we see the line of best fit. Any data points that lie on this line are perfectly predicted values. As we move away from this line, the reliability of the prediction decreases (see Figure 5-7).

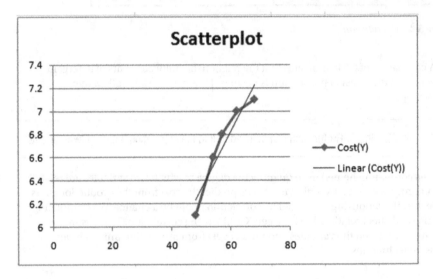

Figure 5-7. *Line of best fit in a scatter plot*

So how do we find the best fit line? The most common and widely used technique is the ordinary least square estimate (OLS).

Working of OLS Regression

The OLS linear regression method is the simplest approach to fit a function to the data. It finds the best-fit line by minimizing the sum of squared errors (SSE) of the data. The SSE is the sum of the deviations of actual values from the mean.

Continuing with our previous example, as we want to determine how cost is affected by the detergent production, the target or independent variable is cost (Y) and the production unit (X) is a dependent variable.

> **STEP 5:** Find the slope. Slope can be calculated using the following equation:

$$\text{Slope}(a) = \frac{\sum (Xi - \bar{X})(Yi - \bar{Y})}{\sum (Xi - \bar{X})^2}$$

(Yi-Ymean)*(Xi - Xmean)	(Xi-Xmean)*(Xi - Xmean)
6.076	96.04
0.456	14.44
-0.064	0.64
1.176	17.64
3.876	104.04

Figure 5-8. *Deviation of X and Y from their mean*

Substituting these values in the slope's equation, we get slope (b) as 0.049484536.

STEP 6: Calculate the intercept. Intercept can be calculated using the following equation:

$$\text{Intercept}(b) = \bar{Y} - Slope \times \bar{X}$$

Substituting the calculated values in the intercept equation gives an intercept of 3.859793814.

Substitute the values of a and b in the regression equation to form the regression formula for predicting the cost.

$$\text{Cost}(Y) = b + aX = 3.8597 + 0.04948 * X$$

Suppose we want to predict the cost of production when the unit of production is 90. We can substitute the value of X as 70 in the preceding equation and calculate the cost to be 7.32.

$$\text{Cost}(Y) = 3.8597 + 0.04948*90 = 3.8597 + 3.46 = 7.32$$

Assumptions of OLS

The simplicity of OLS calculation is overshadowed by several key assumptions. All of these assumptions on data should hold true to reap the benefits of the OLS regression techniques.

1. *Linearity*: The true underlying relationship between X and Y is linear.

2. *Homoscedastic*: The variance of residuals must be constant. The residual is the difference between the observed value and predictive value of the target.

3. *Normality*: The residuals/errors should be normally distributed.

4. *No or little multicollinearity*: The residuals/errors must be independent.

■ **Note** OLS is also affected by the presence of outliers in the data. Outlier treatment is necessary before one proceeds with linear regression modeling using OLS linear regression.

OLS Regression in Oracle Advanced Analytics

The method available in Oracle R Enterprise for performing linear regression using OLS is ore.lm(). For R users, ore.lm() is the Oracle optimized version of the lm() function, which performs calculations on an ORE frame leveraging the Oracle database as the compute engine. The central regression algorithm of all the technology stacks of Oracle Advanced Analytics uses GLM for regression analysis, which I will discuss shortly.

To perform OLS linear regression, boot up the R client and follow these steps.

■ **Note** The airfare dataset can be found in the book's source code folder. This data set has around 18 records with Fare as the target variable and distance as a predictor.

STEP 1: Load the ORE library and connect to the database.

```
library(ORE)
 if (!ore.is.connected()) # Check if client is already connected to R
   ore.connect("dmuser", "orcl","localhost", "sibanjan123", all=TRUE)
```

STEP 2: Read the sample csv file using R's built-in read.csv function.

```
airfare <- read.csv("C:/Users/Admin/Dropbox/analytics_book/chapter-6/data/
airfares.csv")
```

STEP 3: Use the ore.push function to convert the R data frame into a temporary oracle database object.

```
airfare_odf <- ore.push(airfare)
```

STEP 4: Use ore.lm for fitting the data to a linear regression model. The fitting is done in the Oracle Database leveraging database compute engine.

```
oreFit <- ore.lm(Fare ~ ., data = airfare_odf)
```

STEP 5: Interpret the summary statistics of the fitted model (see also Figure 5-9).

```
summary(oreFit)
```

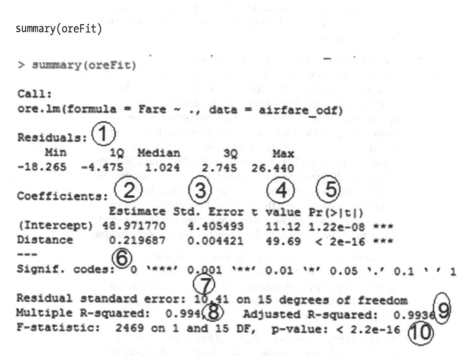

```
> summary(oreFit)

Call:
ore.lm(formula = Fare ~ ., data = airfare_odf)

Residuals: ①
    Min        1Q    Median        3Q       Max
-18.265    -4.475     1.024     2.745    26.440

Coefficients: ②          ③            ④     ⑤
              Estimate Std. Error t value Pr(>|t|)
(Intercept) 48.971770    4.405493   11.12 1.22e-08 ***
Distance     0.219687    0.004421   49.69  < 2e-16 ***
---
Signif. codes: ⑥ 0 '***' 0.001 '**' 0.01 '*' 0.05 '.' 0.1 ' ' 1
                        ⑦
Residual standard error: 10.41 on 15 degrees of freedom    ⑨
Multiple R-squared:  0.994 ⑧    Adjusted R-squared:  0.9936
F-statistic:  2469 on 1 and 15 DF,  p-value: < 2.2e-16 ⑩
```

Figure 5-9. *Linear regression summary results*

1. **Residuals**: Residual is the difference between observed and predicted values of the data. The residual should be normally distributed or close to it when plotted. This means that the average of difference between residual(or leftovers) and actual values is close to zero. As a rule of thumb, the closer the residual is to the normal distribution, the better the model.

2. **Estimate coefficient**: The value of slope and intercept is calculated by the ore.lm function. The intercept in the example is 48.971770, and the slope is 0.219687.

3. **Standard error of coefficient**: This measures the precision of the estimate coefficients and is an estimate of the standard deviation of the coefficients. As a rule of thumb, the lower the standard error of coefficient, the better the estimate. For example, we have two predictors: temperature and pressure, which measures the tensile strength of steel. Our regression model estimates standard error of coefficient for the temperature to be 0.64 and for the pressure as 0.0087. Therefore, our model was able to determine the temperature coefficient with better precision than pressure.

4. ***t-value of the coefficients***: t-value is the ratio of the coefficient estimate over its standard error. It measures the likelihood of a coefficient to be non-zero. A larger t-value indicates that it is less likely that the actual value of coefficient will be zero, and hence it is better.

5. ***p-value of the coefficients***: p-value is used in conjunction with a t-value for hypothesis testing. The null hypothesis for the t-test is that the coefficient has no effect on the model, which means the coefficient is equal to zero. The alternate hypothesis is that the coefficient has some effect on the model. The p-value is the observed significance level for the test. It is compared with the alpha value, which is a chosen significance level. When the p-value is less than the alpha, we accept the alternate hypothesis and reject the null hypothesis. Mostly, alpha is set to 0.05. In this case, if the p-value is less than 0.05, we accept the alternate hypothesis. So a low p-value is considered good, which indicates that the selected predictor is contributing to the predictive power of the model.

6. ***The level of significance***: Significance is computed on the basis on the p-value. It is displayed as stars. As a rule of thumb, the more number of stars, the better the coefficient. A black dot indicates an inadequate coefficient measure and can be discarded from the model.

7. ***Residual standard error / degree of freedom***: This is the standard deviation of the residuals. The closer the residual standard error to zero, the better the model.

8. ***Multiple R-Squared***: Multiple R-squared or the coefficient of determination describes the percentage of variance explained by the predictors for the model. For example, if R^2 is 0.99, it means that the set of predictors used for the model can explain 99% of its variation. As a rule of thumb, the higher the R^2 value, the better the model.

9. ***Adjusted R-Squared***: Multiple R^2 artificially increases when additional predictors are introduced in the model. Adjusted R^2 provides same information as multiple R^2, but it penalizes the metrics whenever any new predictors are added. Thus, this value increases only when the newly added predictor has some effect on the predictability of the model. As a rule of thumb, the higher the adjusted R^2 value, the better the model.

10. ***F-statistics***: The F-statistic performs an F-test on the model. It takes out parameters from the model and checks the model with fewer parameters. If our model performs better than the model with fewer parameters, then the F-statistics will have a higher p-value. As a rule of thumb, a model is considered good when the F-statistics have a higher p-value.

STEP 6: Validate whether the model meets the assumptions that were discussed in the earlier section. Use the plot command to display the following plots, which helps in the validation process.

```
plot(oreFit)
```

The first plot you see is a Residuals vs. Fitted plot, which shows whether the residuals follow a linear pattern. If the residuals are symmetrically distributed around the horizontal dashed line, then they exhibit a perfect linear pattern. For our previous example, the residuals don't follow a linear model (see Figure 5-10).

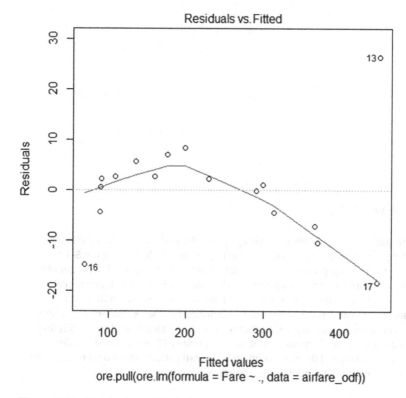

Figure 5-10. Residuals vs. Fitted plot

The second plot is a Normal Q-Q plot, which shows whether the residuals are normally distributed. If they are normally distributed, they are lined up well across the straight dashed line (see Figure 5-11).

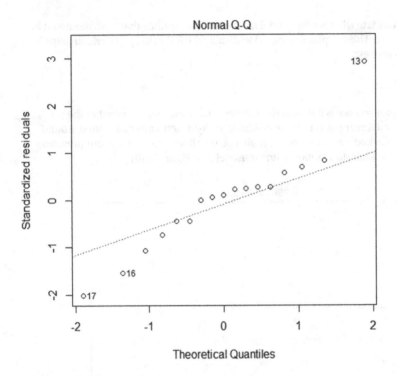

Figure 5-11. *Normal Q-Q plot*

The third plot is the Residuals vs. Leverage plot, which shows whether there are any influential points that affect the slope of the regression model (see Figure 5-12). The influential points (leverage points) appear at the top right corner or top left corner of the graph. These extreme cases are separated from the majority by the Cook's distance(the dashed lines). Cook's distance is useful in finding the data points that influence a regression model. We can treat the data points identified by Cook's distance as outliers and examine them separately. The regression results will be significantly altered if they are excluded from the model. In our model, the data points 17 and 13 are the influential points that affect the slope of the regression line. You can exclude these data points from your analysis and re-run the model to see a better result.

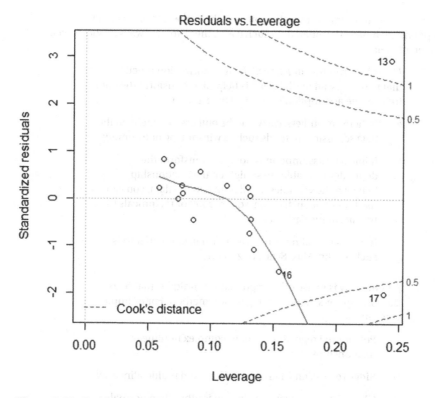

Figure 5-12. *Residuals vs. Leverage plot*

STEP 6: You can use the ore.predict function to predict the airfare for any new distance. It uses the model fit to derive the value for the unknown target.

```
ore.predict(oreFit, airfare_odf)
```

■ **Note**   It is always vital to check two details after building a model: the quality of the model and whether the model meets the assumptions of the algorithm.

As we see in our example, as well as often in the real world data sets, OLS assumptions are mostly invalidated. In such cases, there are two ways to construct a regression model:

1. *Use transformation functions*: The transformation function that we discussed in Chapter 2 is helpful to transform the data to meet the assumption's criteria. For example

    a. If there are outliers, carry out the outliers treatment on the data sets using methods such as winsorizing or trimming.

    b. If linearity assumption is not met, transform the dependent variables to straighten the relationship between the variables. Depending on the data, you can use functions such as log, square root, or reciprocals to transform the variables.

    c. If there are scaling issues, use normalization methods such as Min-Max, Scale, or Z-score.

2. A better way is to use other regression techniques that relax the assumptions of the OLS regression method. For example, you can use

    a. Polynomial regression to fit in complex nonlinear relationships.

    b. Ridge regression when data suffers from multicollinearity.

    c. GLM when the error distribution is other than normal.

I will discuss in detail GLM and ridge regression that are implemented in Oracle Advanced Analytics next.

GLM Regression

GLM extends the family of linear regressions by relaxing the linear relationship and normality assumptions. In GLM, the relationship between predictors and the target need not be linear. For this reason, GLM models can have target variables with distributions other than normal. Using GLM, one can create models for both continuous and categorical target variables having normal, Poisson, and exponential distributions. It can also be used for logistics regression, which is for binary classification problems.

■ **Note**   Logistics regression is a classification technique used for problems where the target has binary values. I will discuss classification in detail in Chapter 6.

GLM has two ingredient functions that help to relax the assumptions of OLS:

- **Link function**: This is used for linearization. The link relates the expected value of the target/response variable to the linear function of the predictors.

- **Variance function**: This expresses variance of the predictors as a function of the target.

Ridge Regression

Ridge regression is used to relax the multicollinearity assumption, that is, when the predictors are highly correlated with each other. It does this by imposing a penalty on the size of the regression coefficients. The ridge regression coefficients try to minimize the penalized sum of square errors. There is a complexity parameter that controls the amount of shrinkage. The higher the value, the more robust coefficients become to collinearity.

Parameters to Tune the GLM Model

Oracle has provided seven parameters to tune the GLM algorithm. It has assigned all parameters with some standard values and can be overridden by specifying the accepted values for each parameter (see Table 5-1):

Table 5-1. *Parameters to Tune the GLM Model in Oracle Advanced Analytics*

Parameter	Default Value	Function It Performs
Generate row diagnostics	Disabled	Generates the model-fitting diagnostics for each row.
Confidence level	0.95	The probability that a certain prediction falls within a range of values; the default is 0.95, which means if repetitive samples are taken and confidence is calculated, then 95% of the interval should contain the population mean.
Missing value treatment	Mean mode	This setting specifies the technique to treat the missing values; the default is the mean mode, which implies the missing numerical values of a variable are to be substituted with its mean, and missing categorical values are to be substituted with its mode. The other option is to delete the row having missing values.
Specify row weights column	Unchecked	Row weight is used to emphasize the values of certain attributes while building the model; by default, it is disabled. If we enable it for an attribute, then the model is biased toward this attribute.

(continued)

161

Table 5-1. (*continued*)

Parameter	Default Value	Function It Performs
Feature selection/ generalization	Unchecked	Feature selection is used to select certain important features/attributes that play an important role in building the model; by default, it is disabled.
Ridge regression	System determined	This option is to enable or disable the ridge regression functionality; by default, it is system determined, which implies the ridge regression will be turned on automatically when the system detects multicollinearity in the data set.
Approximate computation	System determined	By default, it is system determined.

GLM and Ridge Regression in Oracle Advanced Analytics

Unlike OLS regression, which is present only in Oracle R Enterprise, GLM and ridge regression are a part of the Oracle Data Miner as well as Oracle R Enterprise. GLM caters to a wide variety of data sets and hence is the preferred method available in Oracle Advanced Analytics. Ridge regression is available as a property in the algorithm settings. By default, the system determines whether to perform a GLM or a Ridge regression. Ridge regression is turned on if the system detects multicollinearity in the data set.

GLM Regression Using SQL and PLSQL

STEP 1: Create a setting table in the Oracle Database and insert model settings for a GLM. It overrides the default values for the model setting parameters. This table requires at least one record for odms_item_id_column_name to define the item ID parameter. If other records are not inserted, the model-building API proceeds with the default parameters.

The settings table has two attributes:

- setting_name: Name of the setting parameter
- setting_value: Value of the setting parameter

```
set serveroutput on
CREATE TABLE glmr_settings_demo (
  setting_name  VARCHAR2(30),
  setting_value VARCHAR2(4000));
/
-- Turn on feature selection and generation
--
BEGIN
-- Populate settings table
  INSERT INTO glmr_sh_sample_settings (setting_name, setting_value) VALUES
    (dbms_data_mining.algo_name, dbms_data_mining.algo_generalized_linear_
model);
  INSERT INTO  glmr_sh_sample_settings (setting_name, setting_value) VALUES
    (dbms_data_mining.prep_auto, dbms_data_mining.prep_auto_on);
  -- turn on feature selection
  INSERT INTO  glmr_sh_sample_settings (setting_name, setting_value) VALUES
    (dbms_data_mining.glms_ftr_selection,
    dbms_data_mining.glms_ftr_selection_enable);
  -- turn on feature generation
  INSERT INTO  glmr_sh_sample_settings (setting_name, setting_value) VALUES
    (dbms_data_mining.glms_ftr_generation,
    dbms_data_mining.glms_ftr_generation_enable);

  /* Examples of possible overrides are shown below. If the user does not
     override, then relevant settings are determined by the algorithm

  -- specify a row weight column
    (dbms_data_mining.odms_row_weight_column_name,<row_weight_column_name>);
  -- specify a missing value treatment method:
    Default:  replace with mean (numeric features) or
              mode (categorical features)
       or delete the row
    (dbms_data_mining.odms_missing_value_treatment,
     dbms_data_mining.odms_missing_value_delete_row);
  -- turn ridge regression on or off
     By default the system turns it on if there is a multicollinearity
    (dbms_data_mining.glms_ridge_regression,
     dbms_data_mining.glms_ridge_reg_enable);
  */
END;
```

STEP 2: Create the mining model.

To create the regression model, you need to call the CREATE_MODEL Procedure of the DBMS_DATA_MINING package with the parameters in Table 5-2.

Table 5-2. *Parameters for Creating a GLM Model Using DBMS_DATA_MINING.CREATE_MODEL*

Parameter	Mandatory	Description
model_name	Yes	To assign a meaningful model name
mining_function	Yes	The data mining function that is to be used; for regression, you need to call DBMS_DATA_MINING. REGRESSION function
data_table_name	Yes	Name of the data table/view to be used to create the model
case_id_column_name	No	The unique record identifier for the data set; it's mandatory when row level diagnostics is applied
settings_table_name	Yes	Name of the table that contains the model settings values

The syntax for creating a new regression model using PLSQL API is as follows:

```
BEGIN
  DBMS_DATA_MINING.CREATE_MODEL(
    model_name           => 'DEMO_GLMR_MODEL',
    mining_function      => DBMS_DATA_MINING.REGRESSION,
    data_table_name      => 'DEMO_GLMR_AIRLINE_DATA',
    case_id_column_name  => 'CUSTOMER_ID',
           target_column_name   => 'FARE',
    settings_table_name  => 'glmr_settings_demo'
    );
END;
/
```

> **STEP 3:** Check to see if the model is created successfully (see also Figure 5-13).

```
select * from user_mining_models where model_name = 'DEMO_GLMR_MODEL'
```

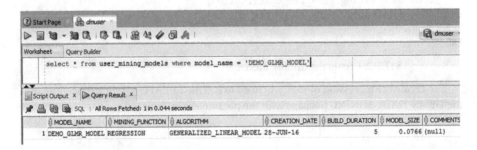

Figure 5-13. *User mining model record*

STEP 4: Check the model settings that were used to create the model (see Figure 5-14)

```
SELECT setting_name, setting_value
  FROM user_mining_model_settings
 WHERE model_name = 'DEMO_GLMR_MODEL'
ORDER BY setting_name;
```

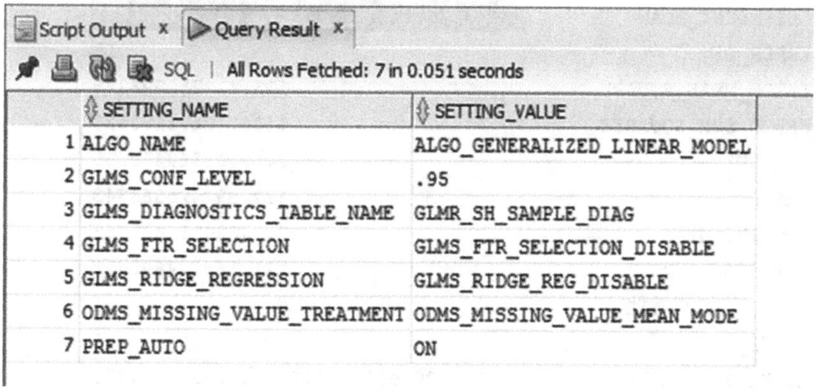

	SETTING_NAME	SETTING_VALUE
1	ALGO_NAME	ALGO_GENERALIZED_LINEAR_MODEL
2	GLMS_CONF_LEVEL	.95
3	GLMS_DIAGNOSTICS_TABLE_NAME	GLMR_SH_SAMPLE_DIAG
4	GLMS_FTR_SELECTION	GLMS_FTR_SELECTION_DISABLE
5	GLMS_RIDGE_REGRESSION	GLMS_RIDGE_REG_DISABLE
6	ODMS_MISSING_VALUE_TREATMENT	ODMS_MISSING_VALUE_MEAN_MODE
7	PREP_AUTO	ON

Script Output × ▷ Query Result ×
SQL   All Rows Fetched: 7 in 0.051 seconds

Figure 5-14. Model settings for the GLM model

STEP 5: To evaluate the model result and its quality, execute the following query. This will return the value of the statistical parameters that are useful in assessing the quality of the model (see Figure 5-15). The method to interpret the important statistical measures such as adjusted R-squared, R-squared, and F-test were discussed in earlier sections.

```
SELECT *
  FROM TABLE(dbms_data_mining.get_model_details_global('DEMO_GLMR_MODEL'))
ORDER BY global_detail_name;
```

165

	⊜ GLOBAL_DETAIL_NAME	⊜ GLOBAL_DETAIL_VALUE
1	ADJUSTED_R_SQUARE	0.993559542604694
2	AIC	81.5334653438706
3	COEFF_VAR	4.55969336760005
4	CORRECTED_TOTAL_DF	16
5	CORRECTED_TOT_SS	269331.882352941
6	DEPENDENT_MEAN	228.352941176471
7	ERROR_DF	15
8	ERROR_MEAN_SQUARE	108.413782093225
9	ERROR_SUM_SQUARES	1626.20673139838
10	F_VALUE	2469.29560479075
11	GMSEP	122.990425063743
12	HOCKING_SP	7.74384157808753
13	J_P	121.168344692428
14	MODEL_CONVERGED	1
15	MODEL_DF	1
16	MODEL_F_P_VALUE	0.00000000000000000461829278964351
17	MODEL_MEAN_SQUARE	267705.675621544
18	MODEL_SUM_SQUARES	267705.675621544
19	NUM_PARAMS	2

Figure 5-15. *Model quality estimates*

Creating GLM Regression Using Oracle R Enterprise

You need to boot up the R client console and use the following steps for creating association rules using Oracle R Enterprise.

STEP 1: Load the ORE library.

```
library(ORE)
```

STEP 2: Use ore.connect to connect to the database.

```
if (!ore.is.connected()) # Check if client is already connected to R
    ore.connect("dmuser", "orcl","localhost", "sibanjan123")
```

STEP 3: Read the CSV file using the read.csv function. This converts the data into a data frame in R.

```
ore.sync("DMUSER","DEMO_GLMR_AIRLINE_DATA",use.keys=TRUE)
airline_df<-ore.get("DEMO_GLMR_AIRLINE_DATA",schema="DMUSER")
```

STEP 4: Create a GLM model using the ore.odmGLM method.

```
oreGLMFit <- ore.odmGLM(FARE ~ ., data = airline_df)
```

STEP 5: Summarize the model fit statistics to validate the quality of the model (see Figure 5-16).

```
summary(oreGLMFit)
```

```
> summary(oreGLMFit)

Call:
ore.odmGLM(formula = FARE ~ ., data = airline_odf)

Residuals:
    Min      1Q  Median      3Q     Max
-16.165  -4.462   1.234   3.873  27.507

Coefficients:
              Estimate Std. Error t value Pr(>|t|)
(Intercept) 51.323316   6.608854   7.766 1.93e-06 ***
DISTANCE     0.219653   0.004538  48.400  < 2e-16 ***
CUSTOMER_ID -0.258175   0.529157  -0.488    0.633
---
Signif. codes:  0 '***' 0.001 '**' 0.01 '*' 0.05 '.' 0.1 ' ' 1

Residual standard error: 10.69 on 14 degrees of freedom
Multiple R-squared:  0.9941,    Adjusted R-squared:  0.9932
F-statistic:  1172 on 2 and 14 DF,  p-value: 2.6e-16
```

Figure 5-16. *GLM regression summary results*

STEP 6: To perform ridge regression, enable it manually by setting the ridge and ridge.vif parameter to TRUE. The VIF is the variation inflation factor, which detects the multicollinearity between the predictors.

```
oreRidgeFit <- ore.odmGLM(FARE ~ ., data = airline_odf, ridge = TRUE,ridge.
vif = TRUE)
```

STEP 7: As in step 5, pull out the model summary statistics to validate the quality of the model (see Figure 5-17).

```
summary(oreRidgeFit)
```

```
> summary(oreRidgeFit)

Call:
ore.odmGLM(formula = FARE ~ ., data = airline_odf, ridge = TRUE,
    ridge.vif = TRUE)

Residuals:
    Min     1Q  Median     3Q     Max
-16.165  -4.462   1.234   3.873  27.507

Coefficients:
             Estimate    VIF
(Intercept)   51.3233  0.000
DISTANCE       0.2197  0.000
CUSTOMER_ID   -0.2582  0.002

Residual standard error: 10.69 on 14 degrees of freedom
Multiple R-squared:  0.9941,     Adjusted R-squared:  0.9932
F-statistic:  1172 on 2 and 14 DF,  p-value: 2.6e-16
```

Figure 5-17. Ridge regression summary results

Creating GLM and Ridge Regression from SQL Developer

Create a data mining project and a workflow. Refer to the steps discussed in Chapter 2 to create a new project and workflow. Once a new workflow is created, follow these steps to create a cluster model in SQL Developer.

> **STEP 1:** Create a data source for table DEMO_GLMR_ AIRLINE_DATA (Figure 5-18). Refer to previous chapters for the details on creating a data source.

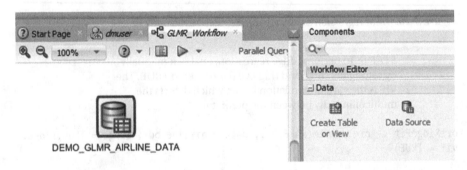

Figure 5-18. Table DEMO_GLMR_AIRLINE_DATA added as a data source node

STEP 2: To build a regression model, select and drag the Regress Build node from the Model section in Components editor to the worksheet. Connect the data source tab and the Assoc Build model as shown in Figure 5-19.

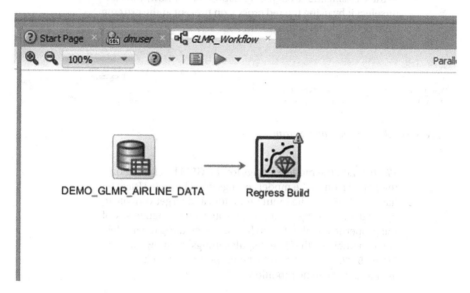

DEMO_GLMR_AIRLINE_DATA   Regress Build

Figure 5-19. *Regress Build node connected to DEMO_GLMR_AIRLINE_DATA node*

STEP 3: Edit the settings of the Regress Build node (see Figure 5-20). Right-click and select Edit Settings from the list. Enter the required details for building a regression model—target and case ID. The target is the independent variable, and case ID is the unique identifier for each record. The processing will be slow if you don't enter the case ID. However, it is mandatory to select the case ID if you want the row diagnostics option to be enabled. Row diagnostics generates the model-fitting diagnostics for each row.

Figure 5-20. *Edit Regression Build Node*

STEP 4: You can see two algorithms available for regression analysis (Figure 5-21). GLM is the one we will be using for this example. SVM is primarily a classification algorithm that can also be used for a regression problem. I will discuss on SVM in the Classifications chapter (Chapter 6). For now, you can unselect it by using the red cross icon present at the extreme right of the Model Settings row.

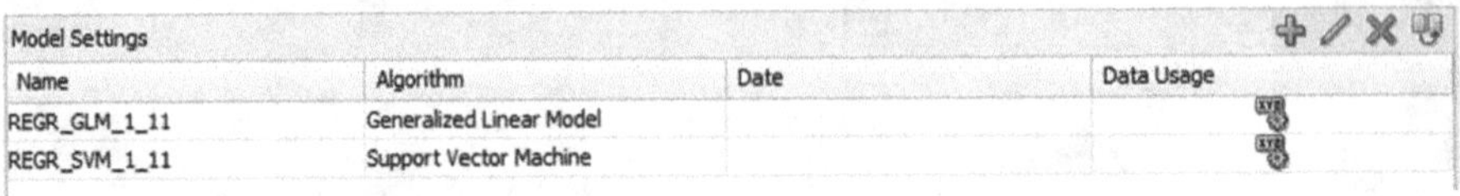

Model Settings				
Name	Algorithm	Date	Data Usage	
REGR_GLM_1_11	Generalized Linear Model			
REGR_SVM_1_11	Support Vector Machine			

Figure 5-21. *Regression Model Settings*

STEP 5: Edit the model settings for the GLM by clicking on the pencil icon in the model settings tab. The Data Usage tab has the details for the input predictors and target variable to be used for building the model. If you want to change any of the properties for the data to be used in creating the model, you can uncheck the "Use default settings" as shown in the Figure 5-22. This will enable the properties for each data attribute that can be modified.

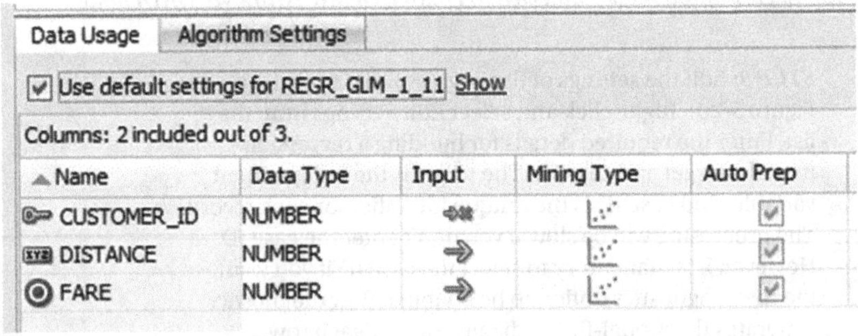

Data Usage	Algorithm Settings			
☑ Use default settings for REGR_GLM_1_11 Show				
Columns: 2 included out of 3.				
▲ Name	Data Type	Input	Mining Type	Auto Prep
☞ CUSTOMER_ID	NUMBER	⇥✖	.:	☑
DISTANCE	NUMBER	⇒	.:	☑
◉ FARE	NUMBER	⇒	.:	☑

Figure 5-22. *Data Usage tab*

STEP 6: Once you are done with validating the data usage for the model, navigate to the Algorithm Setting tab next to it (see Figure 5-23). The default values that were discussed in Table 5-1 are already populated in the algorithm settings. Enter the values for the model settings of your choice to override the defaults.

Data Usage	Algorithm Settings	

The default settings should work well for most use cases. For information on changing model algorithm settings, click Help.

☑ Generate Row Diagnostics

Confidence Level: `0.95`

Missing Value Treatment: `Mean Mode ▼`

☐ Specify Row Weights Column: `CUSTOMER_ID ▼`

☐ Feature Selection `Option...`

☐ Feature Generation `<System determined> ▼`

Ridge Regression `Disable ▼` `Option...`

Approximate Computation: `<System determined> ▼`

Figure 5-23. *Algorithm Settings*

> **STEP 7:** Execute the workflow and wait until it completes successfully (see Figure 5-24).

DEMO_GLMR_AIRLINE_DATA   Regress Build

Figure 5-24. *Final GLM demo workflow*

> **STEP 8:** As discussed in earlier sections, we have to evaluate the performance of a regression model both regarding its quality and accuracy in predicting the results. To measure the quality of the model, select the model from View Model sections. This opens up a new sheet in SQL Developer that has three tabs: (a) Details,(b) Coefficients, and (c) Model Results. The Coefficient tab lists the predictors, their values, and the coefficients generated by the model (see Figure 5-25).

Attribute	Value	Standardized Coeffid...	Coefficient	Standard Error	Wald Chi-Square	Pr > Chi-Square	Lower Coefficient Limit	Upper Coefficient Limit
DISTANCE			0.219655054510971	0.0061	35.9494	0	0.2039	0.2354
<Intercept>		0	46.9053547786379	5.3364	8.7898	0.0003	33.1878	60.6229

Figure 5-25. Model Coefficients

The Details tab lists the model statistics to describe the quality of the model (Figure 5-26).

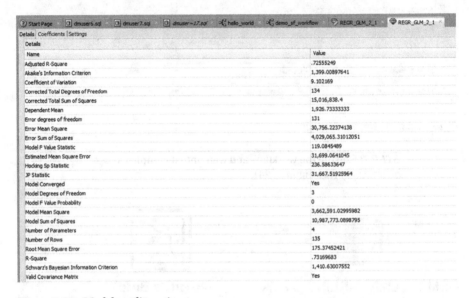

Name	Value
Adjusted R-Square	.72555249
Akaike's Information Criterion	1,399.00897641
Coefficient of Variation	9.102169
Corrected Total Degrees of Freedom	134
Corrected Total Sum of Squares	15,016,838.4
Dependent Mean	1,926.73333333
Error degrees of freedom	131
Error Mean Square	30,756.22374138
Error Sum of Squares	4,029,065.31012051
Model F Value Statistic	119.0845489
Estimated Mean Square Error	31,699.0641045
Hocking Sp Statistic	236.58633647
JP Statistic	31,667.51925964
Model Converged	Yes
Model Degrees of Freedom	3
Model F Value Probability	0
Model Mean Square	3,662,591.02995982
Model Sum of Squares	10,987,773.0898795
Number of Parameters	4
Number of Rows	135
Root Mean Square Error	175.37452421
R-Square	.73169683
Schwarz's Bayesian Information Criterion	1,410.63007552
Valid Covariance Matrix	Yes

Figure 5-26. Model quality estimates

The Settings tab lists the model settings and the input variables used to create the model (see Figure 5-27).

Name	Value
Details │Coefficients │Diagnostics │Settings	
Summary Inputs	
General	
Type	Regression
Owner	DMUSER
Model Name	REGR_GLM_1_11
Target Attribute	FARE
Creation Date	6/28/16
Duration(Minutes)	0.0167
Size(MB)	0.0766
Comment	
Algorithm	
Algorithm Name	Generalized Linear Model
Automatic Preparation	On
Confidence Level	0.95
Enable Ridge Regression	Disable
Feature Selection	Disable
Missing Values Treatment	Mean for Numeric, Mode for Categorial
Row Diagnostics Table Name	ODMR$REGR_GLM_1_11_RD
Variance Inflation Factor	Disable

Figure 5-27. GLM model summary

Once the quality of the model is established, it is important to check the accuracy of the model in terms of its predictive power and performance on the test data set. To view the test results, right-click on the Regress Build model and select the model from the View Test Results option. It pops up a new screen in SQL developer that defaults to the Performance tab.

The Performance tab lists the statistical measures that are used to establish the accuracy of the model (see Figure 5-28). Table 5-3 represents a comprehensive list of the model statistics and their description.

Table 5-3. Description of the Regression Model Statistics

Model Statistics	Description
Predictive Confidence	Measures how much better the predictions are than the average model.
Mean Absolute Error	Measures how close the predictions are to the actuals; it is the average of the absolute values of the errors/residuals.
Root Mean Square Error	Also measures the closeness between the prediction and actuals. However, it emphasizes the large residuals more as it squares them; squaring a large number is even larger than squaring a small number.
Mean Predicted Value	The average of the values predicted by the model on the test data set.
Mean Actual Value	The average of the actual values of the test data set.

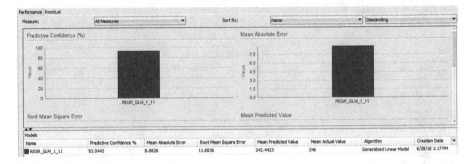

Figure 5-28. *Performance tab*

Residual is the other tab present in the View Test Results option. It displays the residual plot, which helps evaluate the normality of the residuals (Figure 5-29). The closer the residuals to the zero line, the better the prediction result.

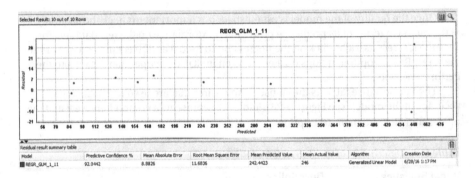

Figure 5-29. *Residual tab*

Guidelines for Regression Modeling

Before we move on to the case study, let's understand the methodology for approaching a regression model problem. This can be applied in general for any supervised learning algorithm.

1. ***Data Preparation***: This is the initial stage in which you explore the data and decide on the target variable and predictors. Preprocess the data as per requirements.

2. ***Split data for cross-validation***: Cross-validation is important to test the model for generalization. Using the whole data set to create a model might not perform well with an unknown or new data set. This phenomenon is known as overfitting. To overcome this, the data is split into a training set and a test set. Training set constitutes 60% of the data, and the rest (40%) is held back for the test set. The model is created on the training data set, and the results are tested on the test data.

3. ***Choose the appropriate supervised learning algorithm***:
   A lot of understanding goes into which algorithms have to be chosen based on the training data currently available. You need to go through the nature of target variable, details of algorithms, and the problem to be solved.

4. ***Fit the model to the training data set***: Apply the algorithm to the training data set to train the model.

5. ***Evaluate the model statistics and tune the parameters***:
   Validate the model based on its statistical parameters, algorithm assumptions, and execution time. If you are not satisfied, tune the model and algorithm settings for better results.

6. ***Evaluate the model on the test data set***: Apply the newly built model to the test data to validate it, see if it performs as expected, and if the results from the test data set are in sync with the training data. This gives you confidence that the model would perform well with the new data when it is deployed for operations.

7. ***Apply the model to a new data set***: If there is not much difference between the results of the training and test data set, deploy the model in production and apply it to the new data set.

Case Study: Sales Forecasting

If you think Sales Forecasting is hard, try running business without a forecast. That's much harder —B-Plans

Sales forecasting is a technique employed by Organizations to predict what future sales will be like. It enables them to take informed business decisions such as what inventory levels to maintain, where and how to allocate resources, and how to identify sales opportunities. Without sales forecasts, organizations will have a hard time predicting their future success and staying competitive in the market.

According to the American Marketing Association, "Sales forecast is an estimate of Sales, in monetary or physical units, for a specified future period under a proposed business plan or program and under an assumed set of economic and other forces outside the unit for which the forecast is made."

Sales forecasts can be developed by utilizing the expertise of experienced personnel that have been in business for years. This is a qualitative method of forecasting sales. Quantitative methods such as time series and regression models are more data driven and robust methods in this dynamic world.

Time series forecast models have a time or temporal component as one of the attributes and sales as the other. They project future sales by considering only the history of past sales figures. There are many methods such as the autoregressive integrated moving average, Holt Winter's Method, and moving averages to do a time series sales forecast. I will discuss moving averages in this case study. Other methods are out of the scope of this book.

Linear regression methods are the preferred technique to understand the underlying factors that influence the sales of an organization. This helps us to prepare a forecast model based on the casual factors that affect the sales of products. Most of the time, both the time series and linear regression techniques are used to design a robust sales forecast model.

In this section, we will create a sales forecast system for a hypothetical superstore BigMart.

Business Understanding

BigMart is the world's leading retailer of an internationally recognized luxury brand in watches, fashion wear, furniture, and appliance stores. You are invited by the CEO of BigMart to develop a sales forecasting system for the future sales of watches. They have experienced low sales for watches in the past year and want to boost it up this year. The forecasting must consider the effect of its sales and the effect of the sales of other products such as fashion wear, furniture, and toys. Also, the CEO wants the forecasting engine to create short-term monthly forecast so that he can control his investment on the watches and determine whether to continue with the product category in future.

Data Understanding

BigMart has a record of the 10-year sales history of different products. You are required to build a model to predict the sales of watches in the coming months based on the previous sales of various products.

Table 5-4 shows the metadata provided to you for your forecasting exercise.

Table 5-4. *Metadata for BigMart's Sales History*

Table	Description
ORDER_DATE	A unique identifier for each record, it aggregates the sales history per month-year
WATCH	Historical sales data for watch products
FASHION_WEAR	Historical sales data for fashion wear products
FURNITURE	Historical sales data for furniture products
ELECTRONICS	Historical sales data for electronics products
JEWELRY	Historical sales data for jewelry products
TOYS	Historical sales data for toy products
MEN_CLOTHING	Historical sales data for men clothing
WOMEN_ CLOTHING	Historical sales data for women clothing
PERSONAL_CARE_ PRODUCTS	Historical sales data for personal care products

Data Preparation

The starting point for your sales forecast is to determine the relevant products that affect the sales of watches. To accomplish that, you need to create a new data mining project and start defining the data sources using your SQL developer. Follow the steps outlined in Chapter 2 to create a new project and a data mining workflow.

Define the BIGMART_SALES_ORDERS data source by dragging the data source node from the Data section of the Components panel and selecting the table BIGMART_SALES_ORDERS. This table contains the data for the metadata described in the Data Understanding section. Then, drag the Filter Columns node from the Transforms panel and connect it to BIGMAR_SALES_ORDER data source (see Figure 5-30).

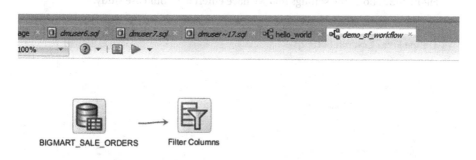

Figure 5-30. *Filter Columns node connected to BIGMART_SALE_ORDERS node*

Edit the Filter Columns node by right-clicking on it and select Edit from the list. This will pop up the list of columns available for the activity. To choose the important attributes that influence the sales of watches, we need to enable the attribute importance functionality. To accomplish that, select Settings present at the top right corner of the Edit Filter Columns Node screen (see Figure 5-31).

Name	Type	Output	Hints
ELECTRONICS	NUMBER	→	
FASHION_WEAR	NUMBER	→	
FURNITURE	NUMBER	→	
JEWELRY	NUMBER	→	
MEN_CLOTHING	NUMBER	→	
ORDER_DATE	VARCHAR2	→	
PERSONAL_CARE_PRODUCTS	NUMBER	→	
TOYS	NUMBER	→	
WATCH	NUMBER	→	
WOMEN_CLOTHING	NUMBER	→	

Figure 5-31. *Edit Filter Columns Node*

This displays another screen to define the settings to filter the columns based on their data. For example, here you can define filter criteria to discard the columns that have unique values less than 95%. This screen also has the property to enable the Attribute Importance functionality. Once it is enabled, you need to select the Target variable, Importance Cutoff, and Top N fields. The attribute importance algorithm evaluates the predictors on their power to predict the target, produces an important score based on this predictive power, and ranks them in decreasing order of their importance. Importance Cutoff and Top N are the criteria to create rules based on your requirement for choosing the variables. For example, if you want to select only those predictors that have predictive power of more than 50%, then you can set the criteria for the Importance Cutoff to be 0.5.

Figure 5-32 shows the settings that we have entered for our case study.

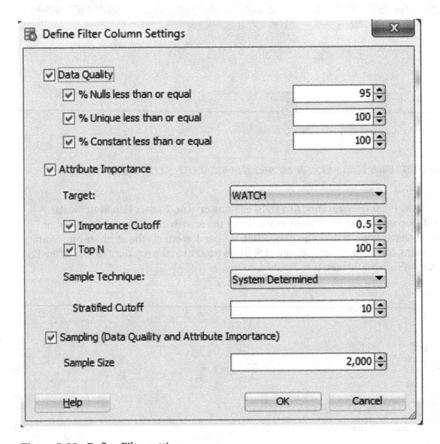

Figure 5-32. *Define Filter settings*

Click on OK and execute the node. Once completed, edit the Filter Columns node again to view the rules generated by the Attribute Importance algorithm. The hints field in the columns section shows the rules that are created for each column. Rank indicates the level of the predictors, and Importance lists the score for each predictor based on their predictive power.

As our intention is only to consider the predictors that have an importance score of more than 0.5, we can exclude other predictors from the list to be used in building the model. To do that, click on the green arrow in the output column for each predictor having a score less than 0.5. This will append another red cross mark to the green arrow, which denotes that the variable is not going to be further used in the current workflow.

For the case study, exclude ELECTRONICS, MEN_CLOTHING, PERSONAL_CARE_ PRODUCTS, and WOMEN_CLOTHING from the list. The final screen after the activity looks like Figure 5-33.

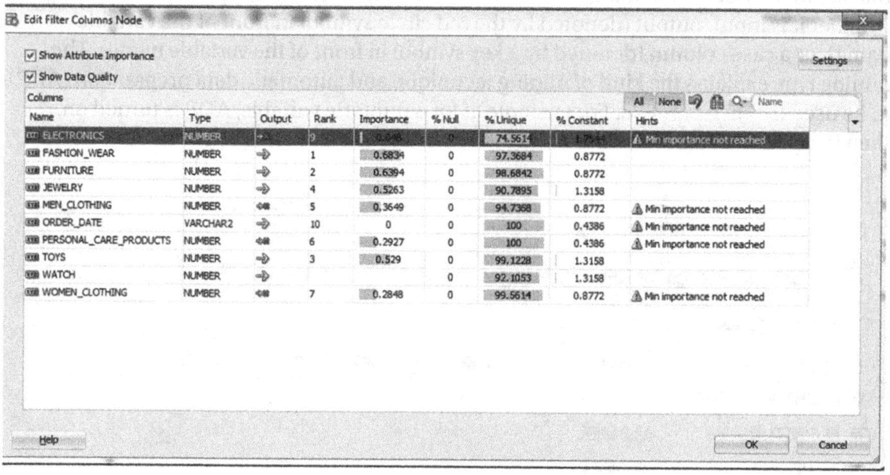

Figure 5-33. *Edit Filter Columns Node after executing the filter node*

Data Modeling

To do the regression analysis, drag the Regression node from the Model section of the Components panel. Connect it with the Filter Columns node and edit its model settings. In the Edit Regression Build node screen (Figure 5-34), you can see three tabs—Build, Input, and Text. The Build section has the information that would be required to build a regression model. Input lists the variables that are to be used for creating the model, and the Text section is for setting up the transformation for the text data. For example, if you want to use customer reviews in a regression model, then you can transform the text sentences for the documents into tokens to be passed as an input to the model.

Exclude the Support Vector Machine from the model settings list by selecting it and then clicking the cross icon.

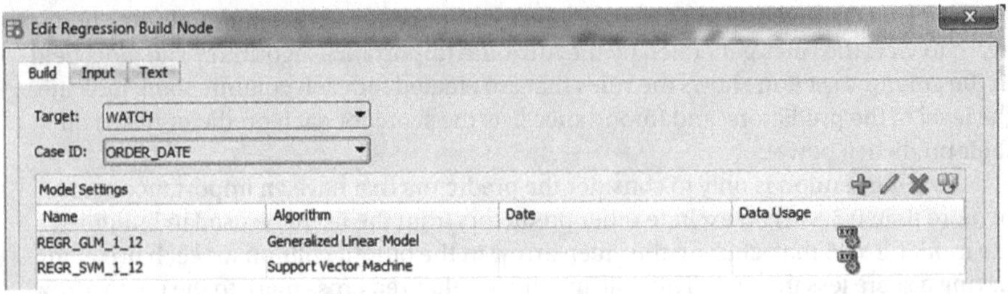

Figure 5-34. *Edit Regression Build Node*

Edit the Model settings for Generalized Linear Model. This screen has two tabs—Data Usage and Algorithm Settings (see Figure 5-35). Data Usage lists all the variables that are to be used for building the model. It also shows the Data Type of the variables, whether it is input, output (denoted by the red circle symbol in front of the variable name), or a case column (denoted by a key symbol in front of the variable name). The mining type explains the kind of mining technique and automatic data preparation (ADP) to be used for each variable. For example, if for a numeric variable, ADP is turned on, then it would normalize its data.

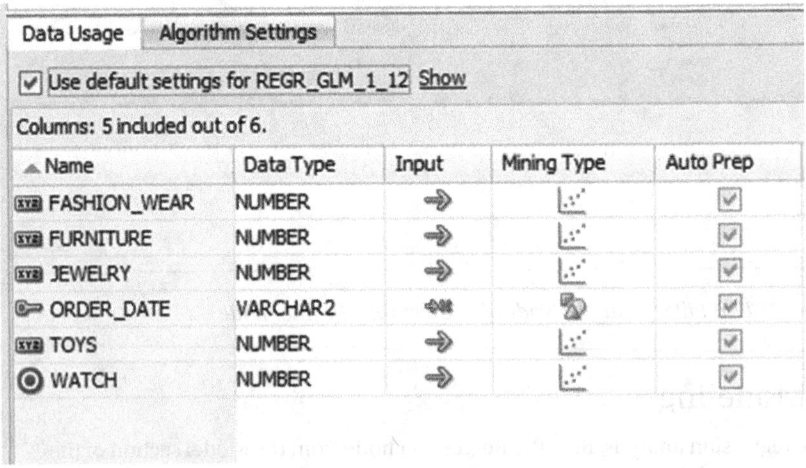

Figure 5-35. *Model Data Usuage tab*

The Algorithm Settings (see Figure 5-36) default to the values that we discussed in the model settings section earlier.

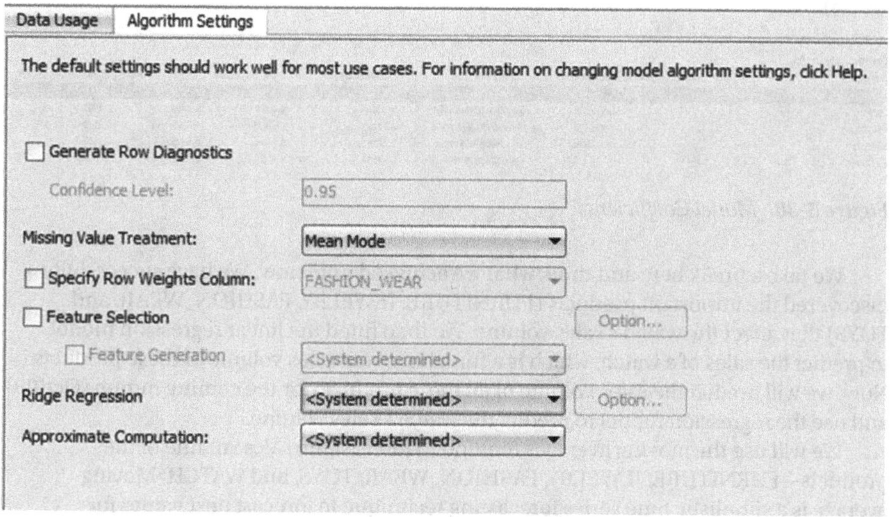

Figure 5-36. *Algorithm Settings*

Once you are happy with the settings to be used by the model, click OK to exit the settings screen and run the Regress Build node. Once it completes successfully (a green tick mark appears at the extreme right corner of the node), evaluate the metrics as discussed in the previous sections. For this model, we have Adjusted R-Square to be .79, which is good, as it explains around 79% of the variance (see Figure 5-37). The predictive confidence is 49.8074%, which is also considered better than average.

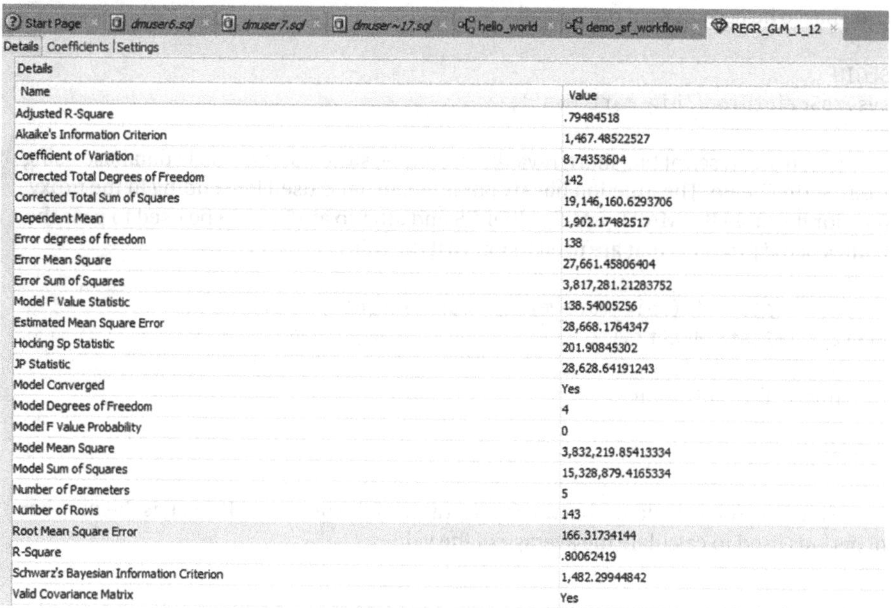

Name	Value
Adjusted R-Square	.79484518
Akaike's Information Criterion	1,467.48522527
Coefficient of Variation	8.74353604
Corrected Total Degrees of Freedom	142
Corrected Total Sum of Squares	19,146,160.6293706
Dependent Mean	1,902.17482517
Error degrees of freedom	138
Error Mean Square	27,661.45806404
Error Sum of Squares	3,817,281.21283752
Model F Value Statistic	138.54005256
Estimated Mean Square Error	28,668.1764347
Hocking Sp Statistic	201.90845302
JP Statistic	28,628.64191243
Model Converged	Yes
Model Degrees of Freedom	4
Model F Value Probability	0
Model Mean Square	3,832,219.85413334
Model Sum of Squares	15,328,879.4165334
Number of Parameters	5
Number of Rows	143
Root Mean Square Error	166.31734144
R-Square	.80062419
Schwarz's Bayesian Information Criterion	1,482.29944842
Valid Covariance Matrix	Yes

Figure 5-37. *Model Quality Estimates*

Attribute	Value	▽ Standardized Coefficient	Coefficient	Standard Error	Wald Chi-Square	Pr > Chi-Square	Lower Coefficient Limit
FASHION_WEAR			0.0488269547800666	0.0231	2.1135	0.0364	0.0031
TOYS		.43614773	0.0778808059524377	0.1286	0.6054	0.5459	-0.1765
FURNITURE		-.26495877	-0.0300139265638928	0.1122	-0.2675	0.7895	-0.2519
JEWELRY		.12412628	0.0744403747649986	0.1219	0.6106	0.5425	-0.1666
<Intercept>		0	919.732672741994	94.5737	9.725	0	732.7318

Details Coefficients Settings
☑ Sort by absolute value   Fetch Size: 1,000 ▷ Query
Coefficients: 5 out of 5   Q▾ Attribute

Figure 5-38. *Model Coefficients*

We take a break here and think what we achieved until now. We have successfully discovered the important products (FURNITURE, JEWELRY, FASHION_WEAR, and TOYS) that affect the watch's sales volume. We then fitted the linear regression model to predict the sales of a watch, which is a function of the sales volume of these products. Now, we will predict the sales volume of all these products for the coming month (Oct-99) and use the regression model to predict the watch's sales volume.

We will use the moving averages method to forecast the sales volume of the products—FURNITURE, JEWELRY, FASHION_WEAR, TOYS, and WATCH. Moving average is a simplistic time series forecasting technique to forecast next values for a set of consecutive values. It is preferable for the short-term forecast, as it averages the successive set values for a set of K time periods. For example, if K is specified as three, then the average is calculated consecutively by considering data for three time periods at a time. As the moving averages forecasting technique is not present in Oracle Advanced Analytics, we will create an R script and interface it to Oracle for in-database computation.

We created the following R script to forecast the sales volume of the products mentioned previously using moving averages.

Drop the R script if already present. For the first-time execution, the sys.rqScriptDrop code has to be commented out from the script.

```
BEGIN
sys.rqScriptDrop('big_mart_mva');
```

Create the R script big_mart_mva that would be stored in the Oracle database using sys.rqScriptCreate. The function has six parameters; dat is used for sending in the proxy data for the table BIGMART_SALE_ORDERS and attr1 to attr5 would be used to pass the names of the products that are used to forecast the sales.

```
sys.rqScriptCreate('big_mart_mva', 'function(dat,attr1,
attr2,attr3,attr4,attr5) {
```

Invoke the ORE library.

```
library(ORE)
```

This function calculates the moving averages of the products. Here, n is the number of periods used to calculate the average of the values.

```
maverage <- function(x,n=3){filter(x,rep(1/n,n), sides=1)}
mv_fit_func applies the maverage function over the products and picks the
last three recent sale values.
mv_fit_func <- function(column){
  tail(maverage(eval((substitute(dat[a], list(a = column)))),3),3)
}
```

In the following, forecast1 to forecast5 variables invoke the mv_fit_func and calculate the average to forecast the sales for next month.

```
forecast1 <- mean(mv_fit_func(attr1))
forecast2 <- mean(mv_fit_func(attr2))
forecast3 <- mean(mv_fit_func(attr3))
forecast4 <- mean(mv_fit_func(attr4))
forecast5 <- mean(mv_fit_func(attr5))
```

The forecast variables are cast to a data frame for the Oracle and R Interface. This interface expects the result to be in a data frame, so that can be stored in the database in a structured format.

```
data.frame(forecast1,forecast2,forecast3,forecast4,forecast5)
}');
END;
/
COMMIT;
```

Once you compile this script in your database, you need to call this script from you data mining workflow. Drag an SQL Query node from the Data section of the Components panel and connect it to the Filter Column node. Double-click on the node to bring up the SQL script editor. This node can be used to invoke any SQL, PLSQL, and R script from the workflow. It has four tabs—Source, Snippets, PL/SQL functions, and R Scripts.

- Source lists the data sources that can be used in the function defined in this node.

- Snippets have the various built-in functions that are available in Oracle.

- PL/SQL functions can be used to invoke any custom PL/SQL package or procedure that we write.

- R scripts nodes are used to invoke R scripts that are stored in the database.

As discussed in earlier chapters, to invoke an R script from SQL or PLSQL, we need to use the R interface functions: rqEval or rqTableEval. The following query is designed using rqTableEval's signature to retrieve the product forecasts from the R script.

```
select to_char(add_months(to_date(max_order_date,'Mon-YY'),1),'Mon-YY')
"ORDER_DATE",rgp.* from (select * from table(rqTableEval(
  cursor(select * from "Filter Columns_N$10002"),
  cursor(select 'FURNITURE' "attr1",
                'JEWELRY' "attr2",
                'FASHION_WEAR' "attr3",
                    'TOYS' "attr4",
                    'WATCH' "attr5",
          1 "ore.connect" from dual),
      'select 1 "FURNITURE_FORECAST",1 "JEWELRY_FORECAST",1 "MEN_CLOTHING_
FORECAST",1 "TOYS",1 "WATCH" from dual',
      'big_mart_mva')) )rgp,
      (select max(ORDER_DATE) max_order_date from
        "Filter Columns_N$10002") filter_columns_tab
```

Copy and paste the preceding script in the R Script tab of the SQL Query Node as shown in Figure 5-39. Click on the green tick symbol present at the top left corner of the script editor to validate the correctness of the script. If correctly validated, the Column section gets populated with the output attribute names. Click OK to exit the editor.

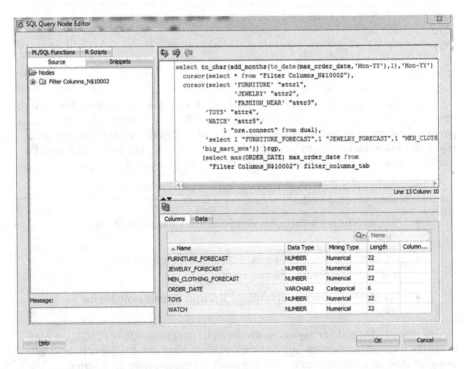

Figure 5-39. *SQL Query Node Editor*

The forecast that is retrieved from the script has to be persisted in a database table to be applied on the linear regression model to predict the sales volume for watches. Drag a Create Table or View node from the Data section of the Components panel and connect it with the SQL Query node. Double-click the node and enter the new table name as PREDICTOR_SALE_FORECAST. Click OK to save the details. This changes the name of the node to PREDICTOR_SALE_FORECAST node. Drag an Apply node from Evaluate and Apply section of the Component panel for applying the regression model to this new data. Connect the Apply node to the Regress Build node and PREDICTOR_SALE_FORECAST node as shown in Figure 5-40.

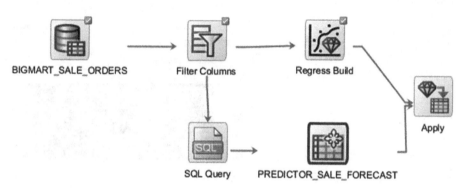

BIGMART_SALE_ORDERS   Filter Columns   Regress Build

Apply

SQL Query   PREDICTOR_SALE_FORECAST

Figure 5-40. *Apply node connected to Regress Build node and PREDICTOR_SALE_FORECAST*

Edit the Apply node to see the results and output columns that would be available as a part of the prediction result set. The Prediction tab lists name of the column that will store the predicted values, lower bound values, and upper bound values of the forecast (see Figure 5-41).

Column	Function	Parameter(s)	Model	Node
REGR_GLM_1_12_PRED	Prediction		REGR_GLM_1_12	Regress Build
REGR_GLM_1_12_PBLW	Prediction Bounds Lower	Confidence (%): 95.0	REGR_GLM_1_12	Regress Build
REGR_GLM_1_12_PBUP	Prediction Bounds Upper	Confidence (%): 95.0	REGR_GLM_1_12	Regress Build

Figure 5-41. *Prediction tab in Apply node*

The Additional Output tab can be used to select additional output columns to be a part of the result set. This is useful to tie back the results with the primary attributes that were used to create the predictions.

Click on the plus icon in the Additional Output tab to select the variables that we want to be the part of our results (see Figure 5-42).

Figure 5-42. *Additional Output tab*

Let's move all the available attributes to the Selected Attributes column so that it helps us to know what values of the predictors influence the sales of the watches (Figure 5-43).

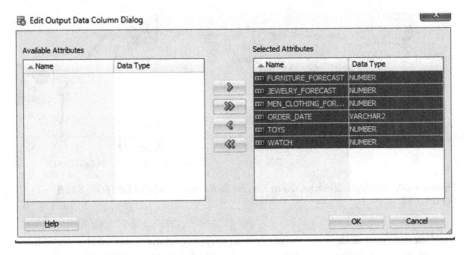

Figure 5-43. *Edit Output Data Column Dialog*

Click OK twice to return to the workflow editor screen. To populate the results in the database, drag the Create Table or View node from the Data section of the component workflow. Enter the name of the output table as BIGMART_WATCH_SALE_FORECAST. The final sales forecast workflow looks like the one in Figure 5-44.

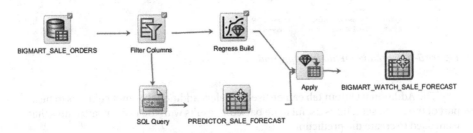

Figure 5-44. *Final sales forecast workflow*

Execute the workflow and wait until it completes successfully. Congratulations!! You have successfully prepared a forecast engine to forecast the sales volume of watches from the forecasts of the products that influences its sales.

Query the BIGMART_WATCH_SALE_FORECAST table to see the forecast results. The forecast for Watches is 2110 for the month of Oct-00, with a 95% confidence interval of 1430 and a 2788 sales volume (see Figure 5-45).

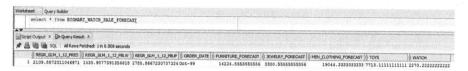

Figure 5-45. Forecast results from BIGMART_WATCH_SALES_FORECAST table

Summary

The purpose of this chapter was to introduce you to Regression analysis, which is a supervised data science technique to predict unknown target labels from known independent attributes. OLS and GLM are widely used algorithms for making predictions with continuous numeric target labels. We discussed forecasting future sales from previous historical sales. We also touched on the process used to identify the factors that affect the sales of a particular product using linear regression. Forecasting is widely used in most of the industries. It might not always be a sales forecast, but it can be used to forecast resource requirements, costs, and cash flows. We successfully prepared a forecasting engine that automates the end-to-end forecasting process. In Chapter 6, we will learn about classification methods, which are used to predict unknown values for discrete or categorical target labels.

CHAPTER 6

■ ■ ■

Classification Methods

In the last chapter, we learned about regression analysis as one of the supervised machine learning methods. It is useful for cases in which we want to predict the future sales of a product, expenditures, or anything that has target labels with continuous numerical data. What happens when the target is a binary, categorical, or discrete variable? Can we use regression analysis for predicting these class labels or do we have any other methods? For example, how would we predict whether a customer is going to churn or not churn, whether a particular promotional activity is going to be successful, or decide whether to sanction a loan to a customer? Regression analysis is not useful in these cases, as they violate linearity assumptions I discussed in Chapter 5. To deal with such situations, we have classification methods. The classification method is another form of supervised machine learning algorithm that is used to predict targets with noncontinuous values. The classification algorithms are known as classifiers, as they identify the set of categories that subsets of data support and use this information to assign a class to an unidentified or unknown target label. Classification finds its applications in various industries such as

- Manufacturing; classifiers can be deployed to predict quality or yield of a product

- Customer relationship functions; classifiers can be used to conduct customer churn analysis for identifying potential customers

- Human resources (HR) functions; classifiers can be deployed to proactively identify high-value employees leaving an organization

- Patient diagnostics; classification can be used to identify medical syndromes based on medical history

- And much more

In this chapter, you will learn about classification methods for data science as per the following outlined topics:

- Overview of classification techniques

- Fundamentals of logistic regression, Naïve Bayes, decision trees, and Support Vector Machines (SVM)

- Some techniques to choose the best classification model

© Sibanjan Das 2016
S. Das, *Data Science Using Oracle Data Miner and Oracle R Enterprise*,
DOI 10.1007/978-1-4842-2614-8_6

- Create classification models using Oracle Advanced Analytics
- Finally, we run through a case study to predict customer churn for a telecommunication organization

Overview of Classification Techniques

Like all supervised methods, classification techniques also require targets with prelabeled classes for training the classifier. For example, if we want to design a classifier for predicting customer churn, then we need past historical data with an indicator to annotate whether the customer churned in the past or not. The classifier for this particular use case can be termed as a "binary" classifier, as it has just two class labels: that is, churned or not churned. When there are more than two class labels, a "multinomial" classifier is used. For example, the quality of a product can be on target, within limits, or above limits. In such cases, we need to prepare a classifier that can identify multiclass targets. However, all classifier algorithms are not one-size-fits-all type. For instance, a binary logistic regression cannot be used for classifying multiclass targets. A multinomial logistic regression can be used, but there other classification algorithms such as SVM and decision trees that can also be used. In the following sections, you will understand the concepts for various classification algorithms and metrics present in the Oracle Advanced Analytics product, which will help to choose the best classifier for our classification task.

Logistic Regression

Logistic regression can be thought of as an extension to linear regression algorithms. It primarily works like that of a linear regression, but it is designed for discrete or categorical outcomes. Logistic regression is used in the case of discrete target variables such as binary responses. In the case of binary responses, some of the assumptions of linear regression given following are not met:

- The target and predictor variables don't follow a linear relationship.
- The error terms are heteroscedastic, which means the size of error varies across the values of predictors.
- The residuals are not normally distributed.

In logistic regression, the target(Y part I discussed in Chapter 5) is transformed to the log of its odds ratio to fit the regression equation as shown here:

$$\log\left(\frac{P}{(P-1)}\right) = a + bX$$

The odds ratio reflects the likelihood or probability of occurrence of a particular event against the likelihood of that same event not taking place. If P is the probability of the occurrence of one event/class, p – 1 is the probability of the occurrence of the second event/class.

You can understand the transformed target $\log\left(\dfrac{P}{(P-1)}\right)$ by taking an example of a

target with two possible outcomes (No or Yes) of customers buying shoes when given a special discount. Let's say, out of 9 customers, 6 brought shoes when given a discount, and 3 did not buy the shoes (see Figure 6-1).

Discount	Buys_shoes
5%	NO
10%	YES
10%	YES
5%	YES
10%	NO
5%	YES
5%	YES
10%	NO
10%	YES

Figure 6-1. Example data

To fit the target attribute to the preceding equation, we need to calculate the odds ratio for both of the result classes (YES/NO).

> **STEP 1:** Calculate the probability(p) for customers who brought shoes, that is, p(Y = Buys_shoes). For example, the probability of customers buying shoes would be the ratio of number of customers who brought shoes to that of the total number of customers, which is 6/9 = 0.66666667 (see Figure 6-2).

P(YES)	0.66666667

Figure 6-2. Probability of customers buying shoes

Similarly, the probability of customers not buying shoes is 3/9
= 0.33333333 as shown in Figure 6-3.

P(NO)	0.33333333

Figure 6-3. Probability of customers not buying shoes

STEP 2: Calculate the odds ratio (p/(p – 1)) for customers
who brought shoes. In our example, p is the probability
of customers who bought shoes; p – 1 is the probability of
customers who didn't buy. Substituting the values from
Figure 6-2 and Figure 6-3 in the odds ratio formula, the odds
of customers buying shoes is (0.66666667/0.33333333), which
equals 2 (see Figure 6-4).

■ **Note** In the case of binary classifications, there are only two choices. If p is the
probability of the first choice, p – 1 is always equal to the probability of the second choice.

Odds(YES)	2

Figure 6-4. Odd of customers buying shoes

STEP 3: Calculate log of Odds(Yes). Log of Odds is also
known as logit. This means $\log\left(\dfrac{P}{(P-1)}\right)$ and logit(Odds(p))
are similar (see Figure 6-5).

log(Odds(YES))	0.30103

Figure 6-5. Log Odds of customers buying shoes

This transformation creates a linear relationship between the target and the
predictors. However, please note that the coefficients are in terms of log odds. For
example, a coefficient of 0.6 implies that a one-unit change in X results in 0.6 unit change
in the log odds of Y.

■ **Note** The Oracle Advanced Analytics methods and code used for GLM regression
in Chapter 5 can be used for performing logistics regression. You can use any data set
provided in this chapter and use those codes to create a logistics regression model.

Naïve Bayes

Naïve Bayes is a classifier based on Bayes' theorem. Before we move on to Bayes' theorem, you need to understand the theory of conditional probabilities that drives Bayes' theorem and Naïve Bayes classifier.

Conditional probability is the probability of something happening given that something else has happened in the past. Here the probabilities of a future event are calculated based on the prior probabilities of some event that has already happened. For example, calculating the financial risk of a potential acquisition based on prior evidences of financial impact that an organization had on previous acquisitions is a possible-use case for Bayes' theorem.

Bayes' theorem or Bayes' rule is a simple mathematical formula for calculating conditional probabilities. Bayes' theorem can be stated as follows:

$$P(T|E) = \frac{P(E|T) * P(T)}{P(E)}$$

In this equation, T is the outcome or the target that we are interested in for prediction. E is the evidence of dependent variables that provides information on classifying the target. P(T) is the overall probability of target classes. P(E) is the overall probabilities of the evidences, and P(E|T) is likelihood of evidences that identifies the classes for outcomes.

Now we try to intuitively understand the basic mechanism of a Naïve Bayes classifier. In this example, we will try to predict whether a customer whom we gave a 5% discount will buy shoes. We have nine records of previous evidences of customers' decisions on buying shoes when they are given a discount of 5% and 10% (see Figure 6-6).

Discount	Buys_shoes
5%	NO
10%	YES
10%	YES
5%	YES
10%	NO
5%	YES
5%	YES
10%	NO
10%	YES

Figure 6-6. *Example data set*

STEP 1: Create a frequency table with the predictors/evidences and outcome as shown in Figure 6-7.

	Buys_shoes	
	YES	NO
5%	3	1
10%	3	2

Figure 6-7. Frequency table

STEP 2: Calculate the overall probabilities of the evidences and outcomes. For example, P(YES) is the number of occurrences of (Buys_shoes = YES) out of the total occurrences of Buys_shoes, which is 6/9 = 0.66666667 (see Figure 6-8).

P(YES)	0.66666667
P(NO)	0.33333333

Figure 6-8. Probabilities of Buy_shoes

Similarly, the overall probabilities of the discount factors 5% and 10% is calculated to be 0.44444444 and 0.55555556, respectively (see Figure 6-9).

P(5%)	0.44444444
P(10%)	0.55555556

Figure 6-9. Probabilities of discount factors

STEP 3: Calculate P(E|T) part of the Naïve Bayes equation. For example, P(5%|YES) is the number of occurrences of (Buys_shoes = YES and Discount = 5%) out of total occurrences of Buys_shoes = YES, which is 3/6 = 0.5 (see Figure 6-10)

P(5% | YES)	0.5
P(5% | NO)	0.33333333
P(10% | YES)	0.5
P(10% | NO)	0.66666667

Figure 6-10. *Prior probalities*

STEP 4: Finally, calculate P(T|E) by substituting these values in the Naïve Bayes equation (see Figure 6-11).

P(YES | 5%)	0.75
P(NO | 5%)	0.44444444

Figure 6-11. *Posterior probabilities*

As the probability of P(YES|5%) is more than P(NO|5%), the prediction of customers buying shoes with an offered discount 5% is more than a 10% discount. Well, this is a hypothetical example. In such a scenario, I would always like to go for buying shoes with 10% discount. If we think in an another direction, the discount might not be the only factor that drives a customer to buy shoes. The shoes associated with 5% discount would be having some other attributes that make them choose these shoes over the shoes with 10% discount. If we have these kinds of scenarios, we might think to pump in more attributes related to shoes and redesign our classifier.

For demonstrating examples using Oracle Advanced Analytics, I will be using HR attrition data, which is present in the accompanying data set folder. It can also be downloaded from the IBM Watson web site. It contains around 1,800 records and denotes whether an employee attrited or not. As this a binary target class, we need to use classification algorithms to prepare the data models.

Creating a Naïve Bayes Model Using SQL and PLSQL

The Naïve Bayes method is the default algorithm for creating classifications using the DBMS_DATA_MINING package. For other algorithms, you have to insert the algorithm name in the setting table. You can follow the steps listed here to create a Naïve Bayes classifier using PLSQL APIs.

STEP 1: Create a table to define the priors for the target classes. The priors represent the distribution of each target class within the data set. By default, the priors are computed from the data distribution. However, there are some cases in which the data is imbalanced, that is, the distribution for one class supersedes the distribution of other classes. In such cases, it requires the priors to be defined for creating a good and unbiased model.

```
CREATE TABLE nb_sample_priors (
  target_value      VARCHAR2(10),
  prior_probability NUMBER);
INSERT INTO nb_sample_priors VALUES ('Yes',0.65);
INSERT INTO nb_sample_priors VALUES ('No',0.35);
```

STEP 2: Create a setting table to store the Naïve Bayes algorithm settings. It overrides the default values for the model-setting parameters.

The settings table has two attributes:

- setting_name: Name of the setting parameter
- setting_value: Value of the setting parameter

```
CREATE TABLE demo_nb_settings (
  setting_name  VARCHAR2(30),
  setting_value VARCHAR2(4000));

BEGIN
  -- Populate settings table NB
  INSERT INTO demo_nb_settings(setting_name, setting_value) VALUES
  (dbms_data_mining.prep_auto,dbms_data_mining.prep_auto_on);
  INSERT INTO demo_nb_settings VALUES
  (dbms_data_mining.clas_priors_table_name, 'nb_sample_priors');
END;
/
```

STEP 3: Create the mining model.

To create the classification model, you need to call the CREATE_MODEL procedure of the DBMS_DATA_MINING package with the parameters shown in Table 6-1.

Table 6-1. *Parameters of the CREATE_MODEL Procedure*

Parameter	Mandatory	Description
model_name	Yes	Assign a meaningful model name for the model
mining_function	Yes	Data mining function that is to be used; for classification, you need to call the DBMS_DATA_MINING. CLASSIFICATION function
data_table_name	Yes	Name of the data table/view to be used to create the model
case_id_column_name	No	Unique record identifier for the data set; if there is no case id, you can assign NULL to this parameter. However, the processing would be slow.
settings_table_name	Yes	Name of the table that contains the model settings values.

The syntax for creating a new regression model using PLSQL API is as follows:

```
BEGIN
  DBMS_DATA_MINING.CREATE_MODEL(
    model_name          => 'demo_class_nb',
    mining_function     => dbms_data_mining.classification,
    data_table_name     => 'HR_EMPLOYEE_ATTRITION',
    case_id_column_name => null,
    target_column_name  => 'attrition',
    settings_table_name => 'demo_nb_settings');
END;
/
```

        STEP 4: Check if the model is created successfully
        (see Figure 6-12).

```
select * from user_mining_models where model_name = 'DEMO_CLASS_NB'
```

MODEL_NAME	MINING_FUNCTION	ALGORITHM	CREATION_DATE	BUILD_DURATION	MODEL_SIZE	COMMENTS
1 DEMO_CLASS_NB	CLASSIFICATION	NAIVE_BAYES	17-JUL-16	2	0.057	(null)

Figure 6-12. *User Mining model record*

STEP 5: Check the model settings that were used to create the model (see Figure 6-13). This helps to keep a track of all the settings used to create a particular model.

```
SELECT setting_name, setting_value FROM user_mining_model_settings
WHERE model_name = 'DEMO_CLASS_NB'
ORDER BY setting_name;
```

⬍ SETTING_NAME	⬍ SETTING_VALUE
1 ALGO_NAME	ALGO_NAIVE_BAYES
2 CLAS_PRIORS_TABLE_NAME	nb_sample_priors
3 NABS_PAIRWISE_THRESHOLD	0
4 NABS_SINGLETON_THRESHOLD	0
5 PREP_AUTO	ON

Figure 6-13. *Model settings for Naïve Bayes*

Once we are done with creating the model, we can validate and check the model quality using various accuracy parameters. The accuracy parameters are explained after we discuss the method to create a classifier using Oracle R Enterprise.

Creating a Naïve Bayes Classifier Using Oracle R Enterprise

You need to boot up the R client console and follow the steps listed here for creating association rules using Oracle R Enterprise.

STEP 1: Load the ORE library.

```
library(ORE)
```

STEP 2: Use ore.connect to connect to the database.

```
if (!ore.is.connected()) # Check if client is already connected to R
    ore.connect("dmuser", "orcl","localhost", "sibanjan123")
```

STEP 3: Use ore.sync to synchronize the metadata for HR_EMPLOYEE_ATTRITION table in the database schema with R environment; and use ore.get to get the proxy ore.frame for this table to assign it to an R variable.

```
ore.sync("DMUSER","HR_EMPLOYEE_ATTRITION",use.keys=TRUE)
attrition_df<-ore.get("HR_EMPLOYEE_ATTRITION",schema="DMUSER")
```

STEP 4: Use ore.odmNB to create a Naïve Bayes model. method.ore.odmNB is the method to call Oracle Data Miner's Naïve Bayes algorithm.

```
ore.nb.mod  <- ore.odmNB(ATTRITION ~ ., attrition_df)
```

STEP 5: Summarize the model statistics. This would show all the prior probabilities for the predictors used in the model (see Figure 6-14).

```
summary(ore.nb.mod)
```

```
Apriori:
      No       Yes
0.825974 0.174026

Tables:
$AGE
    ( ;  33.5),  [33.5;  33.5] (33.5;   )
No                   0.3396226 0.6603774
Yes                  0.6567164 0.3432836

$JOBLEVEL
    ( ;  1.5),  [1.5;  1.5]  (1.5;   )
No                   0.3301887 0.6698113
Yes                  0.6716418 0.3283582

$OVERTIME
              No         Yes
No   0.7610063 0.2389937
Yes 0.4626866 0.5373134

$STOCKOPTIONLEVEL
    ( ;  .5),  [.5;  .5]    (.5;   )
No             0.4088050 0.5911950
Yes            0.7014925 0.2985075

$TOTALWORKINGYEARS
    ( ;  8.5),  [8.5;  8.5]  (8.5;   )
No                   0.3993711 0.6006289
Yes                  0.7014925 0.2985075

$YEARSINCURRENTROLE
    ( ;  4.5),  [4.5;  4.5]  (4.5;   )
No                   0.5974843 0.4025157
Yes                  0.8507463 0.1492537


Levels:
[1] "No"   "Yes"
```

Figure 6-14. Model summary

> **STEP 6:** Use a predict function on a test data set attrition_df to make predictions from the model.

```
# Make predictions
  nb.res  <- predict (ore.nb.mod, attrition_df,"ATTRITION")
```

> **STEP 7:** Compute the confusion matrix for accessing the quality of the model using the following code snippet. The confusion matrix displays the values for actual and predicted attrition of the employees (see Figure 6-15). It helps to calculate many accuracy measures to assess the quality of a model.

We have a detailed discussion on the confusion matrix and other model quality measures in the section "Assessing the model quality for classifiers."

```
with(nb.res, table(ATTRITION,PREDICTION))
```

```
> nb.res  <- predict (ore.nb.mod, attrition_df,"ATTRITION")
> with(nb.res, table(ATTRITION,PREDICTION))
          PREDICTION
ATTRITION  No Yes
      No  293  25
     Yes   36  31
```

Figure 6-15. *Confusion matrix*

Before moving on to other classification methods, you need to understand the measures to assess the quality for a classifier. This will be helpful for you in interpreting and using some of those measures for the Naïve Bayes classifier that we just designed as well as for other classification methods that I will go through in the next sections.

Assessing the Model Quality for Classifiers

The confusion matrix is the primary method used to validate a classifier. Most of the model quality and accuracy metrics are based on the values of the confusion matrix.

1. ***Confusion matrix***: This matrix is a table that contains information about the actual and predicted values for a classifier. It typically looks like Figure 6-16 for a binary classifier.

		Prediction	
		Positive	Negative
Actuals	Postive	a	b
	Negative	c	d

Figure 6-16. *Confusion matrix for a binary classifier*

The data in the confusion matrix have the following meaning:

- "a" is the number of positive class predictions that were correctly identified

- "b" is the number of incorrect predictions for actual positive cases

- "c" is the number of incorrect predictions for negative cases

- "d" is the number of negative class predictions that were correctly identified

2. **Accuracy**: Accuracy measures how often the classifier makes a correct prediction. It is the ratio of the number of correct predictions to the total number of predictions.

$$Accuracy = \frac{a}{a + b + c + d}$$

3. **Precision**: Precision measures the proportions of true positives that were correctly identified.

$$Precision = \frac{a}{a + b}$$

4. **Recall**: Recall is also termed "sensitivity" or "true positive rate." It measures the proportions of true positives out of all observed positive values of a target.

$$Recall = \frac{a}{a + c}$$

5. **Misclassification rate**: It measures how often the classifier has predicted incorrectly.

$$Misclassification\ rate = \frac{c + b}{a + b + c + d}$$

6. **Specificity**: Specificity is also termed a "true negative rate." It measures the proportions of true negatives out of all observed negative values of a target.

$$Specificity = \frac{d}{b + d}$$

7. **ROC (receiver operating characteristic) curve**: The ROC curve is used to summarize the performance of a classifier over all possible thresholds. The graph for ROC curve is plotted with sensitivity in the y axis and (1 - specificity) in the x axis for all possible cutpoints (thresholds).

8. **AUC (area under curve)**: AUC is the area under a ROC curve. If the classifier is excellent, the sensitivity will increase, and the area under the curve will be close to 1. If the classifier is equivalent to random guessing, the sensitivity will increase linearly with the false positive rate(1 - sensitivity). In this case, the AUC will be around 0.5. As a rule of thumb, the better the AUC measure, the better the model.

9. **Lift**: Lift helps measure the marginal improvement in a model's predictive ability over the average response. For example, for a marketing campaign, the average response rate is 5%, but a model identifies segments with a 10% response rate. Then that segment has a lift of 2(10%/5%).

Decision Trees

Decision trees are widely used classifiers in industries for their transparency on describing the rules that lead to a prediction. They are arranged in a hierarchical tree-like structure, simple to understand and interpret. They are not susceptible to outliers. It can be well suited for cases in which we need the ability to explain the reason for a particular decision. For example, sales and marketing departments might need a complete description of rules that influences the acquisition of a customer before they start their campaign activities.

Before moving ahead with the working of a decision tree, let's visualize what a decision tree looks like (Figure 6-17) and the different components associated with a decision tree.

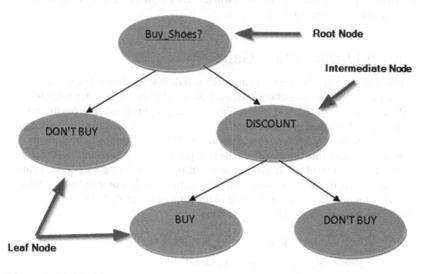

Figure 6-17. Decision tree

The top node of a decision tree is the root node. The nodes at the end of the tree are known as leaf nodes, and the nodes in between the root node and the leaf nodes are the intermediate nodes. The root node is the initial starting point for navigating the branches of the tree. Intermediate nodes derive the path through which we can comprehend the rules for predicting target classes. The navigation to any branch is based on conditions.

Let's take a scenario in which we want to understand the factors that affect customers buying the shoes. As shown in Figure 6-17, we start at the root node and traverse the intermediate nodes that lead us to the leaf node having the target value as BUY. The set of decisions that were made for navigating from the root node to the leaf node are considered as rules to be applied for a particular prediction. However, sometimes this may lead to overfitting the predictions to the data and will have poor performance in the case of new data sets. So, the decision tree can be pruned using some early stopping criteria, which I discuss in the upcoming section "Pruning Methods/Early Stopping Criteria for Decision Trees."

The process of creating a decision tree is a recursive partitioning method where it splits the training data into various groups with an objective to find homogeneous pure subgroups, that is, data with only one class. The splitting criterion used for a decision tree specifies the method to be used for splitting the data. Splitting criterion can be any of the following:

- Minimum number of observations in a node

- Information gain/ entropy-based method—impurity-based criterion that uses entropy as an impurity measure. Entropy is explained in detail in the next section.

- Gini Index—an impurity-based approach that measures the divergences between the probability distributions of the target attribute's values.

Oracle Advanced Analytics provides Gini Index and Entropy-based methods as a splitting criterion. In the following section, I will illustrate how the entropy measure is used by an algorithm to create a decision tree.

Entropy and Information Gain

Entropy is used to measure the degree of impurity in a data set. A data set is impure when it has a mixture of multiple classes. When entropy of a node is zero, it means it contains data for only one class and is pure. Entropy ranges from 0 to 1. It is maximum of 1 when the records are equally distributed among all classes, indicating the maximum level of impurity and the least interesting information. When all records belong to one class, entropy is 0, suggesting the maximum level of purity and the most interesting information. The objective of a decision tree with entropy as a splitting criterion is to find nodes with recursively decreasing entropy until it reaches a node where entropy is zero.

A formal definition of entropy is the sum of the probability of each class label multiplied by the log probability of that same label.

$$\text{Entropy} = -\sum p(x)\log p(x),$$

where p(x) is the probability of an outcome of a target class.

Let's consider a case in which L1 and L2 are two class labels. The entropy equation for this scenario would be as follows:

$$Entropy = -p(L1)\log p(L1) - p(L2)\log p(L2)$$

We can visualize entropy calculations as listed here:

1. **Case 1:** L1 has 6 observations (L1 = 6), and L2 has none (L2 = 0). Using the entropy formula, we can say –6/6*log(6/6) = 0. This means the node has no impurity.

2. **Case 2:** L1 has 5 observations (L1 = 5) and L2 has 1 observation(L2 = 1). Using the entropy formula, we get –1/6*log(1/6) – 5/6 * log(5/6) = 0.65. This means the node has some entropy and can be further split.

3. **Case 3:** L1 has 3 observations(L1 = 3) and L2 has 3 observation (L2 = 3). Using the entropy formula, we get –3/6*log(3/6) – 3/6*log(3/6) = 1. This means the node is entirely impure and is the top criteria to be split.

Information Gain

The decrease in entropy results in information gain. Information gain is the amount of information gained by doing a split for creating a decision tree using the value of a particular attribute. So, if the entropy is minimum, information gain is maximum. A decision tree is constructed by finding values of attributes that return the highest information gain or lowest entropy.

To illustrate the working of a decision tree, consider an example in which we want to classify whether a customer will buy a recently published book. The sample has a target attribute as Decision (Buy/Not Buy) and two predictors: Book Type and Publisher (see Figure 6-18).

Record ID	Book Type	Publisher	Decision(Buy /Not Buy)
BID1	CS	APRESS	BUY
BID2	CS	WILEY	NOT BUY
BID3	BUS	APRESS	BUY
BID4	BUS	WILEY	NOT BUY
BID5	CS	APRESS	BUY

Figure 6-18. Working of a decision tree

STEP 1: Find the attribute that would be a node of the decision tree. The first node is always the root node. To do this, calculate the entropy of the values of the target attribute.

Entropy(T) = -p(Buy)log(p(Buy))-p(Not Buy)log(p(Not Buy))

= -(3/5)log(3/5)-(2/5)log(2/5)

= 0.97095

STEP 2: Next we need to determine the best predictor for the split that would result in maximum information gain. This can be done by subtracting the entropy of each predictor from the entropy calculated in Step 1.

GAIN(T,X) = Entropy(T) - Entropy(T,X)

We have two predictors in our example, so we need to calculate information gain for each of them. The predictor with the highest information gain would be the root node of the tree.

Gain(T,BookType) = Entropy(T)-Entropy(T,CS)-Entropy(T,BUS)
= 0.970951-(2/3)*(-(2/2)*LOG(2/2,2))-(1/2)*LOG(1/2,2))-(1/3)*(-
(1/2)*LOG(1/2,2)-(1/2)*LOG(1/2,2))
= 0.304284
Gain(T,Publisher) = Entropy(T)-Entropy(T,Apress)-Entropy(T,Wiley)
= 0.970951- -(3/5)*(-(3/3)*LOG(3/3,2))-(2/5)*
(-(2/2)*LOG(2/2,2))
= 0.970951-0-0
= 0.970951

STEP 3: Choose the decision node. Decision node is the attribute with the largest information gain. In our example, the information gain is more for the predictor Publisher. So Publisher is selected as the first decision node (root node) for the tree.

STEP 4: Repeat Steps 1 to 3 for each value of the predictors until a branch with an entropy of 0 is found. This node becomes the leaf node, as it has only one class of the target attribute.

From the preceding steps, we saw that the decision tree algorithm natively stops when entropy is zero. However, there are some cases when we want to stop its growth forcibly in between to make it generalized for new data sets. These methods are known as pruning methods.

Pruning Methods/Early Stopping Criteria for Decision Trees

Following are some of the methods that can be used to prune a decision tree for better prediction accuracy:

1. Allow the decision tree to grow fully. Observe the support and confidence of each branch. Create another decision tree by removing the branches with low support and confidence. This can be done by changing the value of the maximum depth parameter of the algorithm with the desired depth. Depth is the length of the longest path from a root node to a leaf node. For example, say a model is prepared with 10 subtrees, and you find the support for the last subtree is very less. You can prepare another decision tree model by setting the parameter maximum depth to be 7.

2. Assign a minimum threshold value for the number of records that can be present in a node. For example, we set the minimum records in a node to 5; then a node can be formed when there are more than 5 records satisfying this rule.

3. Assign minimum records for a split to happen. For example, we set minimum records in a split to be 5; then a node can be further split for achieving purity when the number of records in each split node is more than 5.

All these criterions can be set through the parameters available in Oracle Advanced Analytics for tuning a decision tree model.

Parameters to Tune a Decision Tree Model

Table 6-2 shows the parameters involved in tuning a decision tree.

Table 6-2. Tuning Decision Tree Parameters

Parameter	Default Value	Function It Performs
Homogeneity metric	Gini	Impurity metrics for forming decision trees; it can be either entropy or Gini.
Maximum depth	7	Maximum depth of a tree.
Minimum records in a node	10	Used to set the minimum number of records to be present in a node.
Minimum percent of records in a node	0.05	Same functionality as minimum records in a node; however, the number of records is expressed as a percent of total number of records.
Minimum records for a split	20	Used to set the minimum number of records to be considered for a split.
Minimum percent of records for a split	0.1	Same functionality as Minimum records for a split; however, the number of records is expressed as a percent of total number of records.

Decision Tree Modeling Using SQL and PLSQL

The steps to create a decision tree model are same as other classification models. However, the model-setting table should have a record to indicate the data mining API to use the decision tree algorithm. In this section, I demonstrate steps to create a decision tree model using primary settings and parameters.

> ***STEP 1:*** Create a setting table and insert model settings for a GLM. It overrides the default values for the model-setting parameters. This table requires at least one record for odms_item_id_column_name to define the item id parameter. If other records are not inserted, the model building API proceeds with the default parameters.

The settings table has two attributes:

- setting_name: Name of the setting parameter

- setting_value: Value of the setting parameter

```
CREATE TABLE demo_dt_settings (
  setting_name   VARCHAR2(30),
  setting_value VARCHAR2(4000));

BEGIN
  -- Populate settings table
  INSERT INTO demo_dt_settings VALUES
    (dbms_data_mining.algo_name, dbms_data_mining.algo_decision_tree);
  INSERT INTO demo_dt_settings VALUES
    (dbms_data_mining.tree_impurity_metric, 'TREE_IMPURITY_ENTROPY');

END;
/
```

> ***STEP 2:*** Create the mining model.

To create the decision tree model, you need to call the CREATE_MODEL Procedure of the DBMS_DATA_MINING package with the parameters shown in Table 6-2.

The syntax for creating a new classification model using PLSQL API is as follows:

```
BEGIN
  DBMS_DATA_MINING.CREATE_MODEL(
    model_name          => 'demo_class_dt',
    mining_function     => dbms_data_mining.classification,
    data_table_name     => 'HR_EMPLOYEE_ATTRITION',
    case_id_column_name => null,
    target_column_name  => 'attrition',
    settings_table_name => 'demo_dt_settings');
END;
/
```

STEP 3: Check if the model is created successfully (see Figure 6-19).

```
select * from user_mining_models where model_name = 'DEMO_CLASS_DT'
```

	MODEL_NAME	MINING_FUNCTION	ALGORITHM	CREATION_DATE	BUILD_DURATION	MODEL_SIZE	COMMENTS
1	DEMO_CLASS_DT	CLASSIFICATION	DECISION_TREE	20-JUL-16	3	0.0628	(null)

Figure 6-19. *User mining model record*

STEP 4: Check the model settings that were used to create the model (see Figure 6-20).

```
SELECT setting_name, setting_value
  FROM user_mining_model_settings
 WHERE model_name = 'DEMO_CLASS_DT'
ORDER BY setting_name;
```

	SETTING_NAME	SETTING_VALUE
1	ALGO_NAME	ALGO_DECISION_TREE
2	PREP_AUTO	OFF
3	TREE_IMPURITY_METRIC	TREE_IMPURITY_ENTROPY
4	TREE_TERM_MAX_DEPTH	7
5	TREE_TERM_MINPCT_NODE	.05
6	TREE_TERM_MINPCT_SPLIT	.1
7	TREE_TERM_MINREC_NODE	10
8	TREE_TERM_MINREC_SPLIT	20

Figure 6-20. *Model settings for decision trees*

STEP 5: You can extract the rules for the decision tree model by executing the following query. The result will be XML type. Click on the pencil icon next to the Query Results in SQL Developer to view the XML (see Figure 6-21).

```
SELECT  dbms_data_mining.get_model_details_xml('demo_class_dt')  AS DT_
DETAILS from dual;
```

209

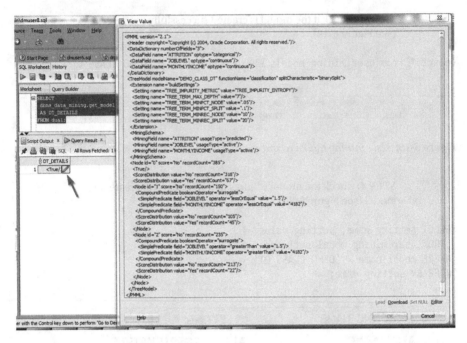

Figure 6-21. *Decision tree rules XML*

Decision Tree Modeling Using Oracle R Enterprise

You need to boot up the R client console and follow the next steps for creating decision trees using Oracle R Enterprise.

> ***STEP 1:*** Load the ORE library.

```
library(ORE)
```

> ***STEP 2:*** Use ore.connect to connect to the database.

```
if (!ore.is.connected()) # Check if client is already connected to R
   ore.connect("dmuser", "orcl","localhost", "sibanjan123")
```

> ***STEP 3:*** Use ore.sync to synchronize the metadata for the HR_
> EMPLOYEE_ATTRITION table in the database schema with
> R environment, and use ore.get to get the proxy ore.frame for
> this table for assigning it to an R variable.

```
ore.sync("DMUSER","HR_EMPLOYEE_ATTRITION",use.keys=TRUE)
emp_df<-ore.get("HR_EMPLOYEE_ATTRITION",schema="DMUSER")
```

210

STEP 4: Create a decision tree (DT) model using the ore.
odmDT method.

```
dt.mod  <- ore.odmDT(ATTRITION ~ ., data= emp_df)
```

STEP 5: Summarize the model fit statistics to see the rules
generated by the decision tree model (see Figure 6-22).

```
summary(dt.mod)


Call:
ore.odmDT(formula = ATTRITION ~ ., data = emp_df)

  n =  385

Nodes:
  parent node.id row.count prediction        split              surrogate
1    NA       0      385       No            <NA>                    <NA>
2     0       1      150       No  (JOBLEVEL <= 1.5) (MONTHLYINCOME <= 4041)
3     0       2      235       No  (JOBLEVEL > 1.5)  (MONTHLYINCOME > 4041)
          full.splits
1             <NA>
2 (JOBLEVEL <= 1.5)
3   (JOBLEVEL > 1.5)

Settings:
                          value
prep.auto                    on
impurity.metric   impurity.gini
term.max.depth                7
term.minpct.node           0.05
term.minpct.split           0.1
term.minrec.node             10
term.minrec.split            20
```

Figure 6-22. *Decision tree model summary*

SVM

SVM is a supervised machine learning algorithm used primarily for classifications.
However, it can also be used for regression problems. It is based on the concept of
decision planes where it divides the data set based on decision boundaries. It finds the
optimal classes by defining a hyperplane that best separates the data. Wikipedia defines
hyperplane to be "a subspace of one dimension less than its ambient plane." This means
that if we have a two-dimensional data set, then the hyperplane is a one-dimensional
plane, that is, a line. If we use a data set that is three-dimensional, then hyperplane will be
a two-dimensional plane.

Without getting too much into the mathematics of SVM, you can understand
intuitively how SVM works. The SVM modeling option will do the math for you. However,
you need to understand the concept and parameters that are essential to execute an SVM
algorithm.

Assume you have a data set of examination results for five students. The results are based on two subjects—History and Geography. Our objective is to prepare an SVM classifier that can classify the results into Pass/Fail category for other unclassified students.

History	Geography	Result
84	49	Pass
69	33	Fail
67	37	Fail
78	45	Pass
81	49	Pass

Figure 6-23. *Examination results table*

From the table in Figure 6-23, we can see that there are only two predictors: History and Geography. The target for the classifier is results. So clearly this is a two-dimensional data set that is linearly separable using a line. We can plot the data points on a graph as shown in Figure 6-24.

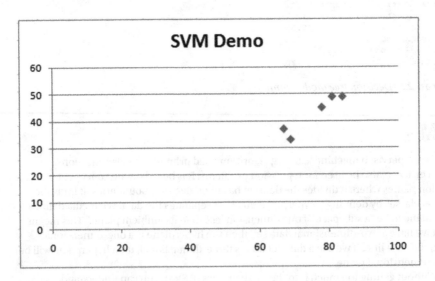

Figure 6-24. *Examination result plot*

The first question an SVM classifier asks is whether the data set is linearly separable, and there is a way to divide the data points into two distinct classes using a line.

As we can see from Figure 6-25, there can be multiple lines that separate the data points into two distinct classes.

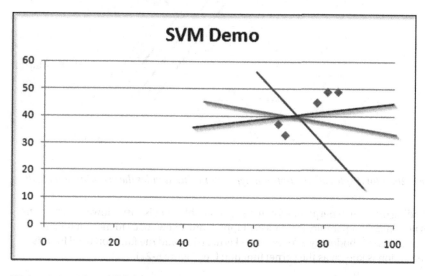

Figure 6-25. *Plot of different lines that can separate examination results data*

So the next question the SVM classifier asks, Is there a way to optimally divide the data points using a line? An optimal line (or hyperplane) is one that is as far as possible from the data points of each target class.

To find an optimal hyperplane, the distance between the hyperplane and the closest data point is computed. Once we have this value, it is doubled to get the margin. An optimal hyperplane is one with the biggest margin. Therefore, the final objective of the SVM can be defined as finding the optimal separating hyperplane that maximizes the margin of the training data (see Figure 6-26).

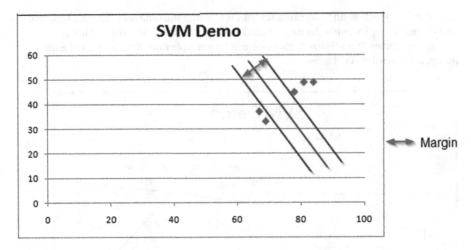

Figure 6-26. *Plot displaying the optimal hyperplane that divides the examination result*

Finding optimal hyperplanes for linearly separable data is easy. However, when the data is not linearly separable, the data is mapped into a new space to make it linearly separable. This methodology is known as a kernel trick, and the function used for this transformation is known as the kernel function (see Figure 6-27).

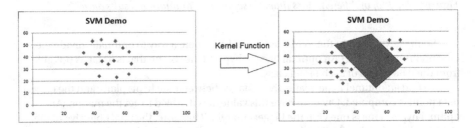

Figure 6-27. *Visualization of the kernel function*

Parameters to Tune the SVM Model

Oracle Advanced Analytics provides different parameters to tune the SVM algorithm, which are shown in Table 6-3 and described in the following sections.

Table 6-3. *Parameters to Tune the SVM Model*

Parameter	Default Value	Function It Performs
Kernel function	System determined	By default, the algorithm determines which kernel function to use for the model; available kernel methods are linear and Gaussian.
Tolerance value	0.001	This is the model convergence criteria; a larger tolerance value will lead to the build model faster, but it is less accurate.
Specify the complexity factor	Unchecked	By default; you need not specify any complexity factor. If you specify a complexity factor, it should be a positive number; the complexity factor is responsible for avoiding overfitting and underfitting a data set to a model.
Active learning	Checked	This restricts the SVM algorithm to use the most informative samples of the data rather than attempting to use the whole data set.

SVM Using SQL and PLSQL

The steps to create an SVM model are the same as other classification models. However, the model-setting table should have a record to indicate the data mining API to use in the SVM algorithm. In this section, we demonstrate steps to create a SVM model using primary settings and parameters.

> **STEP 1:** Create a setting table and insert model settings for the SVM model as we did for previous classification models.

The settings table has two attributes:

- setting_name: Name of the setting parameter
- setting_value: Value of the setting parameter

```
set serveroutput on
CREATE TABLE demo_svm_settings (
  setting_name  VARCHAR2(30),
  setting_value VARCHAR2(4000));
/
--
```

```
BEGIN
  -- Populate settings table for SVM
-- Populate settings table
  INSERT INTO demo_svm_settings(setting_name, setting_value) VALUES
  (dbms_data_mining.algo_name, dbms_data_mining.algo_support_vector_
machines);
  INSERT INTO demo_svm_settings (setting_name, setting_value) VALUES
  (dbms_data_mining.svms_kernel_function, dbms_data_mining.svms_linear);
  insert into demo_svm_settings (setting_name, setting_value) VALUES
  (dbms_data_mining.prep_auto, dbms_data_mining.prep_auto_on);

END;
/
```

STEP 2: Create the mining model.

To create the SVM model, you need to call the CREATE_MODEL Procedure of DBMS_DATA_MINING package with the parameters discussed in Table 6-3.

The syntax for creating the new SVM model using PLSQL API is as follows:

```
BEGIN
  DBMS_DATA_MINING.CREATE_MODEL(
    model_name          => 'demo_class_svm',
    mining_function     => dbms_data_mining.classification,
    data_table_name     => 'HR_EMPLOYEE_ATTRITION',
    case_id_column_name => null,
    target_column_name  => 'attrition',
    settings_table_name => 'demo_svm_settings');
END;
/
```

> **STEP 3:** Check to see if the model is created successfully (see Figure 6-28).

```
select * from user_mining_models where model_name = 'DEMO_CLASS_SVM'
```

MODEL_NAME	MINING_FUNCTION	ALGORITHM	CREATION_DATE	BUILD_DURATION	MODEL_SIZE	COMMENTS
1 DEMO_CLASS_SVM	CLASSIFICATION	SUPPORT_VECTOR_MACHINES	20-JUL-16	2	0.0771	(null)

Figure 6-28. *User mining model record*

> **STEP 4:** Check the model settings that were used to create the model (see Figure 6-29).

```
SELECT setting_name, setting_value
  FROM user_mining_model_settings
 WHERE model_name = 'DEMO_GLMR_MODEL'
ORDER BY setting_name;
```

216

	SETTING_NAME	SETTING_VALUE
1	ALGO_NAME	ALGO_SUPPORT_VECTOR_MACHINES
2	PREP_AUTO	ON
3	SVMS_ACTIVE_LEARNING	SVMS_AL_ENABLE
4	SVMS_COMPLEXITY_FACTOR	0.297024
5	SVMS_CONV_TOLERANCE	.001
6	SVMS_KERNEL_FUNCTION	SVMS_LINEAR

Figure 6-29. Model settings for SVM

> **STEP 5:** You can view the results for the SVM model by executing the following query. The result is a table of many complex data types (see Figure 6-30).

```
SELECT details_tab.class, attributes_tab.attribute_name aname, attributes_
tab.attribute_value aval, attributes_tab.coefficient coeff
  FROM TABLE(DBMS_DATA_MINING.GET_MODEL_DETAILS_SVM('DEMO_CLASS_SVM'))
details_tab,
      TABLE(details_tab.attribute_set) attributes_tab
ORDER BY details_tab.class, ABS(attributes_tab.coefficient) DESC;
```

	CLASS	ANAME	AVAL	COEFF
1	Yes	AGE	(null)	-0.99258795541538702
2	Yes	JOBINVOLVEMENT	(null)	-0.66226064212041302
3	Yes	NUMCOMPANIESWORKED	(null)	0.58883702684391903
4	Yes	STOCKOPTIONLEVEL	(null)	-0.55294031950400502
5	Yes	JOBROLE	Manufacturing Director	-0.53102147920639498
6	Yes	JOBSATISFACTION	(null)	-0.50622663082639796
7	Yes	BUSINESSTRAVEL	Non-Travel	-0.49010776690154101
8	Yes	WORKLIFEBALANCE	(null)	-0.48963507223481501
9	Yes	DISTANCEFROMHOME	(null)	0.47972612744170201
10	Yes	ENVIRONMENTSATISFACTION	(null)	-0.471713744114856
11	Yes	BUSINESSTRAVEL	Travel_Frequently	0.39121539324841498
12	Yes	EDUCATIONFIELD	Technical Degree	0.38225694939399202
13	Yes	OVERTIME	No	-0.37058182945411999
14	Yes	OVERTIME	Yes	0.37058182945411999
15	Yes	JOBROLE	Laboratory Technician	0.36707721644860303

Figure 6-30. Model results

SVM Using Oracle R Enterprise

You need to boot up the R client console and follow these next steps for creating an SVM model using Oracle R Enterprise.

STEP 1: Load the ORE library.

```
library(ORE)
```

STEP 2: Use ore.connect to connect to the database.

```
if (!ore.is.connected()) # Check if client is already connected to R
    ore.connect("dmuser", "orcl","localhost", "sibanjan123")
```

STEP 3: Use ore.sync to synchronize the metadata for the HR_EMPLOYEE_ATTRITION table in the database schema with R environment, and use ore.get to get the proxy ore.frame for this table to assign it to an R variable.

```
ore.sync("DMUSER","HR_EMPLOYEE_ATTRITION",use.keys=TRUE)
emp_df<-ore.get("HR_EMPLOYEE_ATTRITION",schema="DMUSER")
```

STEP 4: Create the SVM model using the ore.odmSVM method.

```
ore_svm.mod  <- ore.odmSVM(ATTRITION ~ ., data=emp_df,"classification",kernel.function="linear")
```

STEP 5: Summarize the model fit statistics to see the model coefficient estimates and the important values of the attributes used for building the model (see Figure 6-31).

```
summary(ore_svm.mod)
```

```
Settings:
                        value
prep.auto                  on
active.learning    al.enable
complexity.factor   0.297024
conv.tolerance         1e-04
kernel.function       linear
```

```
Coefficients:
      class              variable                          value       estimate
1     Yes                    AGE                                   -9.932895e-01
2     Yes           BUSINESSTRAVEL                 Non-Travel -4.907448e-01
3     Yes           BUSINESSTRAVEL       Travel_Frequently  3.915672e-01
4     Yes           BUSINESSTRAVEL           Travel_Rarely  9.917761e-02
5     Yes               DAILYRATE                            3.433388e-02
6     Yes              DEPARTMENT          Human Resources  3.346426e-02
7     Yes              DEPARTMENT   Research & Development  2.471865e-02
8     Yes              DEPARTMENT                   Sales -5.818291e-02
9     Yes        DISTANCEFROMHOME                            4.790902e-01
10    Yes               EDUCATION                            1.620376e-01
11    Yes          EDUCATIONFIELD          Human Resources  1.600444e-02
12    Yes          EDUCATIONFIELD            Life Sciences -1.051578e-01
13    Yes          EDUCATIONFIELD                Marketing  3.959439e-02
14    Yes          EDUCATIONFIELD                  Medical -1.936601e-01
15    Yes          EDUCATIONFIELD                    Other -1.386217e-01
16    Yes          EDUCATIONFIELD         Technical Degree  3.818408e-01
17    Yes          EMPLOYEENUMBER                            1.557342e-01
18    Yes   ENVIRONMENTSATISFACTION                        -4.715671e-01
19    Yes                  GENDER                   Female -1.176262e-01
20    Yes                  GENDER                     Male  1.176262e-01
21    Yes               HOURLYRATE                          -9.504074e-02
22    Yes           JOBINVOLVEMENT                          -6.620735e-01
23    Yes                 JOBLEVEL                          -1.755337e-01
24    Yes                  JOBROLE Healthcare Representative -2.687532e-01
25    Yes                  JOBROLE          Human Resources  3.346426e-02
26    Yes                  JOBROLE     Laboratory Technician  3.666566e-01
```

Figure 6-31. *Model results*

A Few Other Important PLSQL APIs for Classifiers

To compute some of the model metrics, Oracle has provided PLSQL APIs shown in Table 6-4. One can use these APIs directly to compute the metrics and store them in tables.

Table 6-4. *PLSQL APIs to Compute Model Accuracy Metrics*

PLSQL API	Description
DBMS_DATA_MINING.COMPUTE_CONFUSION_MATRIX	Computes the confusion matrix
DBMS_DATA_MINING.COMPUTE_LIFT	Computes lift
DBMS_DATA_MINING.COMPUTE_ROC	Computes the ROC values; the values can be used to plot the ROC curve

Oracle has also provided a cool feature known as prediction details, which can be used to see details for a particular prediction. This feature is not available in most of the analytics and data science tools. Most of the time, you want to view the details of a particular prediction such as why an employee has been predicted to attrite by the model. Traditionally, this is possible for decision trees where we can understand the reasons for a prediction by navigating through the branches of a tree. However, in Oracle Advanced Analytics, one can see these details for any model using the PREDICTION_DETAILS API. Therefore, if you want to understand the reason why an employee has been flagged as a candidate for possible attrition, you can view the details by using this API. The following SQL query uses the prediction_details API to extract the details for each prediction.

```
select
      prediction(DEMO_CLASS_SVM using *) pred_value,
      prediction_probability(DEMO_CLASS_SVM using *) pred_prob,
      prediction_details(DEMO_CLASS_SVM using *) pred_details
from HR_EMPLOYEE_ATTRITION;
```

After you run this query, each prediction record has the pred_details XML results (see Figure 6-32). When the XML file is expanded, you can view the rules and ranks of each attribute that lead to a particular prediction.

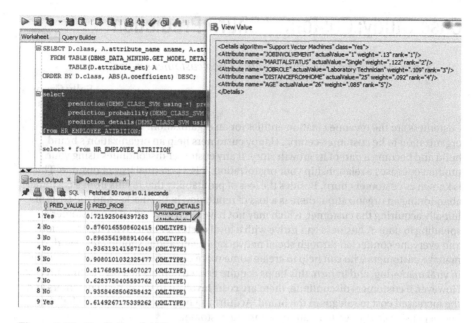

Figure 6-32. Prediction details output

Choosing a Classifier

In some applications, the accuracy of a prediction is the only thing that matters. It may not be important to know how the model works. In others, the ability to explain the reason for a decision can be crucial. The choice of classification algorithm also can be based on the size of the data set and features that you have. If you only have a few samples, you might not use a complex classifier that will overfit on your data.

The commonly used technique is to try creating a bunch of models using different algorithms from sample data. Select the best model by comparing the model metrics such as accuracy measures and AUC. The higher the AUC for a model, the better it performs on that particular data set than other models; and thus, it can be used for that task. Although the accuracy of the algorithm is important, sometimes meeting an accuracy threshold is all that is required. Most important are the values in the quadrants of a confusion matrix and the context in which you are making the predictions. For example, too many false negatives in case of a fraud detection task are not acceptable; whereas in the event of medicine innovation, higher false positives are not acceptable.

The speed of building models and predicting outcomes is also of vital importance depending on the environment where the predictions are to be done. In cases where predictions are to be done in real time, accuracy can be compromised for performance.

Case Study: Customer Churn Prediction

"Good customer service costs less than bad customer service."

—Sally Gronow

Customers are the revenue creation entities for an organization. This calls for an organization to be customer-centric. Happy customers help an organization's brand build and become a part of its growth story. If any customer discontinues using your product or ceases a relationship your organization, it is a potential loss of revenue. This is known as customer churn. Besides the loss of profitability that results from a customer abandoning an organization, there is a loss of reputation in the market and the costs of initially acquiring the customer, which may not have been covered by the customer's spending to date. A business can thrive with a loyal customer base. With digitalization and everyone connected through social networks, organizations should create and manage customers who can help to create some more customers. Loyal customers help in viral marketing, and in turn, this helps acquire new customers at a very minimum cost. However, if customers discontinue, there are costs involved to target new customers and the increased cost to safeguard the brand. Acquiring a new customer is more challenging and expensive compared to retaining a current customer.

Customer Churn prediction is a most important tool for an organization's CRM (customer relationship management) toolkit. Doing it correctly helps an organization to retain customers who are at a high risk of churning. Organizations can design marketing actions and campaigns to retain these customers proactively, which contributes to eliminating the risk of customer churn.

In this case study, we will create a churn prediction model for a telecommunications company I call IndTel.

Business Understanding

IndTel has noticed a high rate of churn in recent years. The organization has developed a retention campaign, and the customer analytics team has been advised to identify the customers who have a high probability of churning. These targeted customers will then be placed in the retention campaign program to contribute to the organization's growth story.

Data Understanding

IndTel has provided you the data of past customer churn history for predicting the customers who are going to churn in the future. The metadata is as shown in Table 6-5.

Table 6-5. *Metadata of Historical Customer Churn Data Set*

Data	Meaning
CUSTOMERID	Unique identifier for a customer
GENDER	Gender (male/female)
SENIORCITIZEN	Is the customer a senior citizen?
PARTNER	Is the customer a partner of the organization?
DEPENDENTS	Number of dependents
TENURE	Number of months with the organization
PHONESERVICE	Are they using phone service?
MULTIPLELINES	Do they have multiple connections?
INTERNETSERVICE	Are they using Internet service?
ONLINESECURITY	Have they subscribed for online security?
ONLINEBACKUP	Do they use an online backup facility?
DEVICEPROTECTION	Do they have a device protection plan?
TECHSUPPORT	Did they call tech support any time?
STREAMINGTV	Did they subscribe for streaming TV?
STREAMINGMOVIES	Did they subscribe for streaming movies?
CONTRACT	Nature of the contract
PAPERLESSBILLING	Have they opted for paperless billing?
PAYMENTMETHOD	The type of payment method they have chosen
MONTHLYCHARGES	Monthly charges
TOTALCHARGES	Total charges
CHURN	Did they churn (Yes/No)?

Data Preparation

In this case study, we start by randomly sampling the churn data into two sets: Train and Test. Then, we first determine the relevant attributes that affect customer churn. This would help us to reduce the number of attributes to be used by the model. The attributes that don't contribute to the predictive power of the model are not useful; and having those as inputs to the model might increase its processing time unnecessarily. To accomplish that, you need to create a new data mining project and define the data source using your SQL Developer. Follow the steps outlined in Chapter 2 to create a new project and a data mining workflow.

Define the DEMO_CUST_CHURN_TAB data source by dragging the data source node from the data section of the components panel and selecting the table DEMO_CUST_CHURN_TAB. This table contains the data for the metadata described in the "Data Understanding" section. Then drag the Sample node from the Transforms panel to the workflow editor and connect it to the DEMO_CUST_CHURN_TAB data source. Edit the properties of the Sample node and rename it to Train (Figure 6-33).

DEMO_CUST_CHURN_TAB   **Train**

Figure 6-33. *Data mining workflow*

Similarly, connect another Sample node to the DEMO_CUST_CHURN_TAB data source and rename it to Test (Figure 6-34).

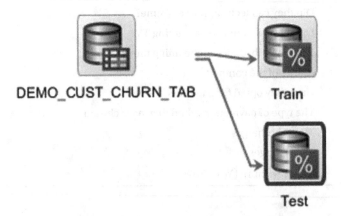

DEMO_CUST_CHURN_TAB   **Train**

                                 Test

Figure 6-34. *Workflow with Train and Test sample nodes*

Right-click on the Test node and edit its settings. By default, the Percent of sample size is 60. Change it to 40. You can have this as any number. But as standard practice, we kept the Train sample as 60% and the Test as 40% of the data set. Select a CASE ID, which helps the sample node to uniquely identify each record (see Figure 6-35).

Edit Sample Node ✕

Sample Size:	Percent ▼
Percent (40%):	⬤
	0 10 20 30 40 50 60 70 80 90 100
Sample Type:	Random ▼
Seed:	12345
Case ID:	CUSTOMERID ▼
	(Supply Seed and Case ID for Reproducibility)

Help   OK   Cancel

Figure 6-35. Test node settings

Click OK to return back to the workflow editor page.

Now, drag a Filter node to the workflow editor and connect it to the Train node as shown in Figure 6-36.

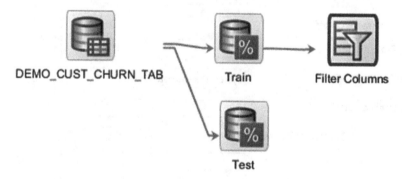

DEMO_CUST_CHURN_TAB   Train   Filter Columns

Test

Figure 6-36. Workflow connecting Filter Columns node

Edit the Filter Columns node by right-clicking on it and select Edit from the list. This will pop up the list of columns available for the activity. To choose the important attributes that influence customer churn, you need to enable the attribute importance functionality. You can follow the detailed procedure described in Chapter 5 to enable the attribute importance functionality. Figure 6-37 is a quick recap of this step. Select Settings at the top right corner of the Edit Filter Columns Node screen.

Figure 6-37. Filter column editor

Enable the attribute importance functionality, select the target attribute, and define the important cutoff value as shown in Figure 6-38.

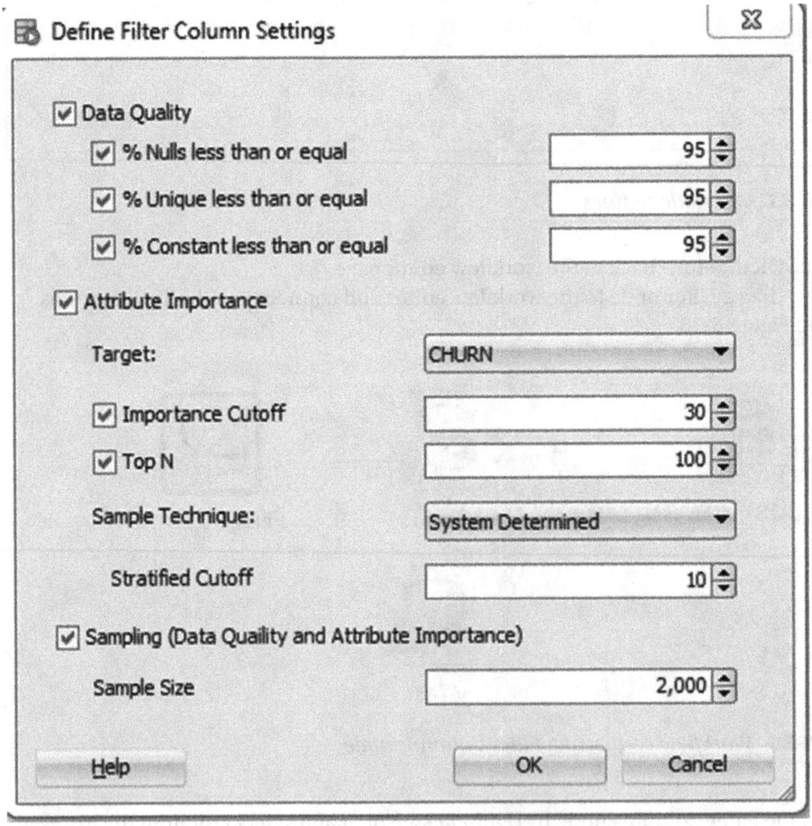

Figure 6-38. Filter Column Settings

Click on OK and run the node. Once completed, edit the Filter Columns node again to view the rules generated by the attribute importance algorithm. As our intention is only to consider the predictors that have importance scores of more than 0.3, we can exclude other predictors from the list to be used in building the model. For the case study, exclude GENDER, MULTIPLELINES, PHONESERVICE, and TOTALCHARGES from the list. The final screen after the activity looks like Figure 6-39.

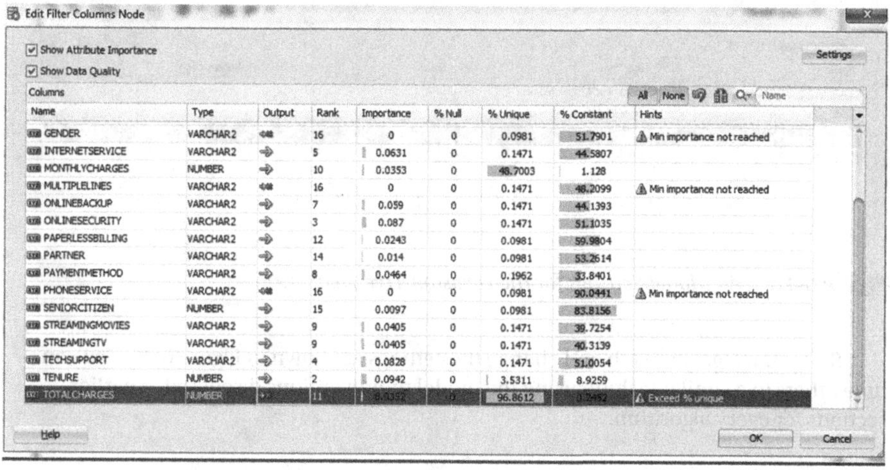

Figure 6-39. *Filter columns with excluded attributes*

Data Modeling

To prepare a classification model, drag the Classification node from the Model section of the components panel. Connect it with the Filter Columns node and edit its model settings. Select Target attribute as CHURN, which is the label of past customer churn records. Also, select Case ID as Customer ID as a unique record identifier (see Figure 6-40). We will prepare this model using the four different classification algorithms and then choose the results of the best classifier to predict the customer churn.

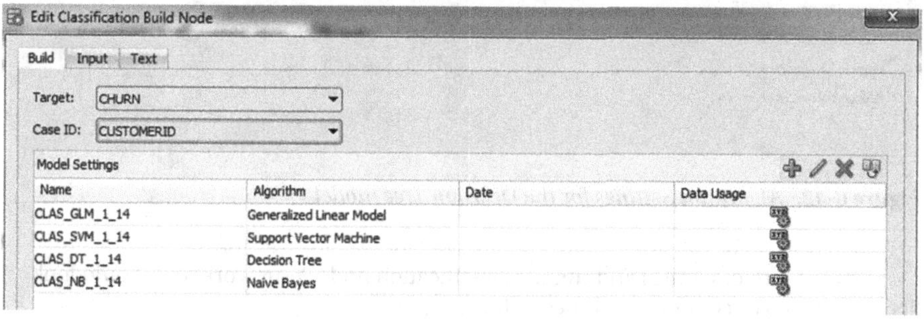

Figure 6-40. *Classification Build Node settings*

If you want to change the algorithm setting of any classifier, select the algorithm and click on the pencil icon as shown in Figure 6-41.

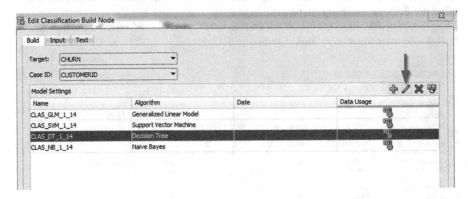

Figure 6-41. *Edit Model Settings for the Decision Tree algorithm*

Select the algorithm tab and change the settings as shown in Figure 6-42. You can tune others in a similar fashion using the model tuning options discussed in earlier sections for every algorithm.

Model Settings					
Name	Algorithm	Date	Data Usage	Columns Excluded by Rules	
CLAS_GLM_1_14	Generalized Linear Model			0	
CLAS_SVM_1_14	Support Vector Machine			0	
CLAS_DT_1_14	Decision Tree			0	
CLAS_NB_1_14	Naive Bayes			0	

Data Usage **Algorithm Settings** Performance Settings

The default settings should work well for most use cases. For information on changing model algorithm settings, click Help.

Homogeneity Metric:	Gini
Maximum Depth:	7
Minimum Records in a Node:	10
Minimum Percent of Records in a Node:	0.05
Minimum Records for a Split:	20
Minimum Percent of Records for a Split:	0.1

Figure 6-42. *Algorithm Settings for the Decision Tree model*

Exit the model settings and run the classification node. Your workflow should look like Figure 6-43 once completed successfully.

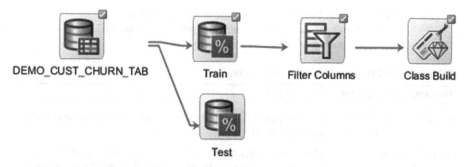

DEMO_CUST_CHURN_TAB   Train   Filter Columns   Class Build

Test

Figure 6-43. *Model workflow*

Selecting the Best Classifier

You will compare different model accuracy measures and select the results of the best classifier for your task. To accomplish this, right-click on the Class Build node and select Compare Test Results shown in Figure 6-44.

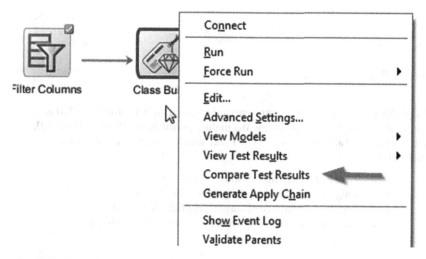

Figure 6-44. *Compare Test Results*

The Compare Test Results window has five tabs: Performance, Performance Matrix, ROC, Lift, and Profit. All these tabs show the comparison using different metrics of all the algorithms used in the Class Build model.

> *Performance*: This screen shows the comparison of various
> performance measures for different algorithms used for the
> model creation (see Table 6-6 and Figure 6-45).

Table 6-6. *Performance Measures Description*

Performance Measures	Description
Predictive Confidence	This measures how good the predictions are compared to the average model
Overall Accuracy (%)	This is the percentage of correct predictions made by the model when compared with the actual target classes of the test data set
Average Accuracy (%)	This is the average per-class accuracy percentage of correct predictions made by the model when compared with the actual target classes of the test data set

Name	Predictive Confidence %	Overall Accuracy %	Average Accuracy %	Cost	Algorithm
■ CLAS_DT_1_14	46.9487	75.6789	73.4743	918.7253	Decision Tree
■ CLAS_GLM_1_14	49.2952	73.1442	74.6476		Generalized Linear Model
■ CLAS_NB_1_14	44.4037	67.1696	72.2019		Naive Bayes
□ CLAS_SVM_1_14	48.7924	72.6011	74.3962		Support Vector Machine

Figure 6-45. *Performance measures*

The performance measures for our case study seem to indicate that the GLM is
best among all four algorithms, providing the highest predictive confidence (49.2952),
second highest overall accuracy percent (73.1442), and highest average accuracy
percent(74.6476).

> *Performance Matrix*: This tab shows the comparison of
> different algorithms based on the percentage of correct
> predictions, the percentage of correct predictions for each
> target class, and the count of correct predictions (Figure 6-46).

Display:	Compare Models ▼				
Models	**Correct Predictions %**		**Correct Predictions Count**	**Total Case Count**	**Total Cost**
CLAS_GLM_1_14	73.1442		1,212	1,657	0
CLAS_SVM_1_14	72.6011		1,203	1,657	0
CLAS_DT_1_14	75.6789		1,254	1,657	918.7253
CLAS_NB_1_14	67.1696		1,113	1,657	0

Figure 6-46. *Performance matrix*

Also, one can view the confusion matrix for each classifier by changing the value for Display list to "Show Detail" as shown in Figure 6-47.

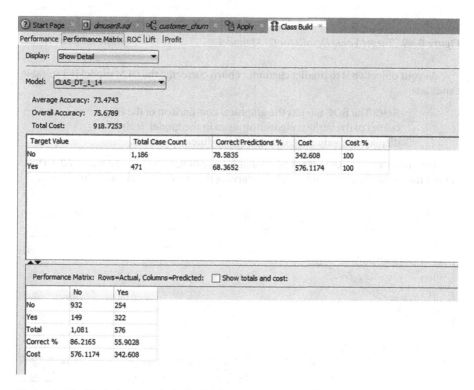

Figure 6-47. *Confusion matrix in the Performance Matrix tab*

The performance matrix for our case study seems to indicate DT (Decision Tree) as the best among all four algorithms providing the highest percentage of correct predictions. However, when we check the target details, we can see that the true positives (number of cases in which customer churn is Yes) for DT is 68.3652 (see Figure 6-48).

Target Value Details		
Measure	Correct Predictions	
▲ Target Value	CLAS_DT_1_14	
No		78.5835
Yes		68.3652

Figure 6-48. *Target Value Details for DT model*

Whereas for the GLM, the number of true positives is 78.1316 (see Figure 6-49).

Target Value Details		
Measure	Correct Predictions ▼	
Target Value	CLAS_GLM_1_14	
No		71.1636
Yes		78.1316

Figure 6-49. *Target Value Details for GLM model*

As your objective is to predict customer churn correctly, the GLM model is a suitable candidate.

> *ROC*: The ROC tab has the graphical comparison of the ROC
> curves of the various algorithms used in the model. It also
> displays the values for the AUC of the different models.

For our case study, the graphical comparison of ROC curves is a bit cluttered, so we cannot make any finite decision just by observing the graphs (see Figure 6-50).

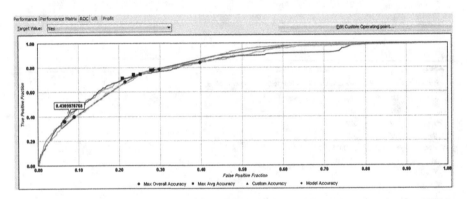

Figure 6-50. *ROC curve for all classification models*

If we look at the AUC values, the GLM is best among all four algorithms providing the highest AUC (0.8304; see Figure 6-51).

Models		
Model	Area Under Curve	Max Overall Accuracy %
■ CLAS_DT_1_14	0.8162	76.4635
■ CLAS_GLM_1_14	0.8304	78.6964
■ CLAS_NB_1_14	0.8096	78.3343
■ CLAS_SVM_1_14	0.8088	77.1273

Figure 6-51. *Area Under Curve values*

Lift: The Lift tab has a graphical comparison of the Lift measures of the various algorithms used in the model along with the values for Lift and Gain measures. For our case study, the Lift curves are very closely tied (see Figure 6-52), so we need to check the numeric values of the Lift.

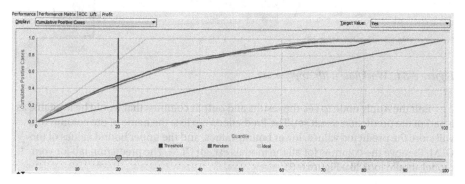

Figure 6-52. *Lift chart*

We can see the Lift Cumulative value for GLM is 2.3074, which is the highest among all other algorithms (see Figure 6-53).

Models						
Model	Lift Cumulative	Gain Cumulative %	Records Cumulative %	Target Density Cumulative	Algorithm	Creation Date
CLAS_DT_1_14	2.1669	44.463	20.519	0.6159	Decision Tree	7/21/16 2:34 AM
CLAS_GLM_1_14	2.3074	47.3461	20.519	0.6559	Generalized Linear Model	7/21/16 2:34 AM
CLAS_NB_1_14	2.2971	47.1338	20.519	0.6529	Naïve Bayes	7/21/16 2:34 AM
CLAS_SVM_1_14	2.1315	43.7367	20.519	0.6059	Support Vector Machine	7/21/16 2:34 AM

Figure 6-53. *Lift values*

We can conclude that the GLM is performing well for predicting the customer churn for IndTel. We will use the results of the GLM model to predict customer churn for the holdout data set (test data) and store the results in IndTel's customer database.

Drag an Apply node from the Evaluate and Apply section of the component panel for applying the Class Build model to the Test data set. Connect the Apply node to the Class Build node and Test node as shown in Figure 6-54.

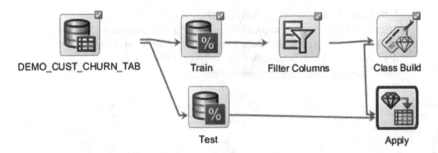

Figure 6-54. *Workflow with Apply node*

Edit the Apply node to see the results and output columns that would be available as a part of the prediction result set. The Predictions tab lists the name of the column that will store the predicted values, lower bound values, and the upper bound values of the GLM. Remove the columns for all the models except GLM. The prediction tab for your model should look like Figure 6-55.

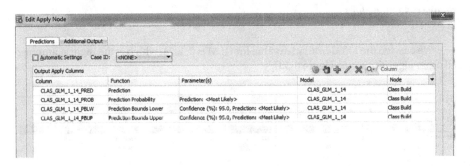

Figure 6-55. *Apply Node editor*

In the Additional Output tab, select CUSTOMERID to be stored as an attribute along with the predictions in the output table (see Figure 6-56). This will help to tie the predictions to the customer master records of IndTel.

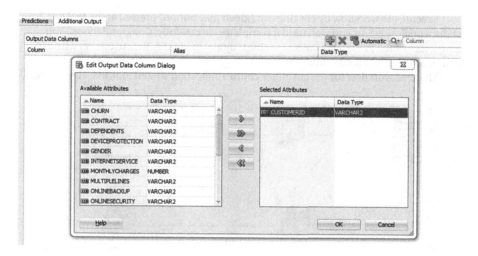

Figure 6-56. *CUSTOMERID as an additional attribute*

Drag the Edit Create Table or View Node from the Data section of the component workflow to store the prediction results in the database. Edit the Create Table or View Node to enter the result table name as DEMO_CUST_CHURN_TEST as shown in Figure 6-57.

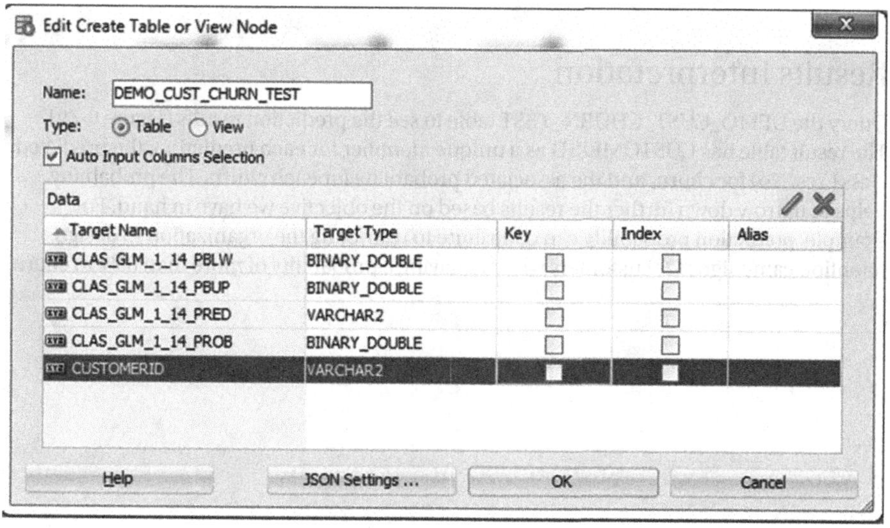

Figure 6-57. *Edit Create Table or View Node*

235

Execute the workflow by clicking run (green play button) present at the tool bar of the workflow editor and wait until it completes successfully. Congratulations!! You have successfully prepared a customer churn model to identify the customers who can potentially churn in the future.

The final workflow for the customer churn application should look like Figure 6-58.

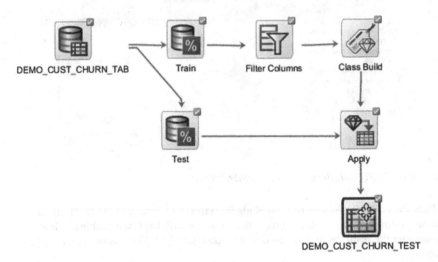

Figure 6-58. *Final workflow*

Results Interpretation

Query the DEMO_CUST_CHURN_TEST table to see the prediction results (Figure 6-59). The result table has CUSTOMERID as a unique identifier for each prediction, the predicted class (Yes/No) for churn, and the associated probability for each churn. The probability helps to narrow down further the results based on the objective we have in hand. For example, prediction probability can contribute to restricting the organization to design retention campaigns for customers who are having a probability of more than 80% to churn.

Worksheet	Query Builder

`select CUSTOMERID, CLAS_GLM_1_14_PRED, CLAS_GLM_1_14_PROB from DEMO_CUST_CHURN_TEST`

▶ Query Result ×

SQL | Fetched 50 rows in 0.01 seconds

	CUSTOMERID	CLAS_GLM_1_14_PRED	CLAS_GLM_1_14_PROB
1	1017-FBQMM	No	0.6841437281118994
2	2305-MRGLV	Yes	0.5484924756501286
3	9209-NWPGU	No	0.8011967734349948
4	6461-PPAXN	No	0.8119092888283229
5	9350-ZXYJC	No	0.8091751650668828
6	6582-OIVSP	Yes	0.6216253263147113
7	3786-WOVKF	No	0.6417132125401045
8	2809-ILCYT	No	0.8384060950111442
9	3387-VAIUS	Yes	0.8795713472160984
10	4612-THJBS	Yes	0.5240929870632124
11	0137-UDEUO	No	0.7075203618130118
12	3583-EKAPL	Yes	0.7649248624025823
13	8382-SHQEH	Yes	0.850592608149209
14	6284-AHOOQ	Yes	0.754108107144161
15	2342-CKIAO	No	0.7014473727248982
16	8375-KVTHK	No	0.6342927480915094
17	3001-CBHLQ	Yes	0.7427592677663695
18	3077-RSNTJ	No	0.7695624057396822
19	9208-OLGAQ	Yes	0.8201522999641767

Figure 6-59. Customer churn prediction results

Summary

In this chapter, I introduced you to classification methods, which are a form of supervised learning to predict the target with discrete and noncontinuous class labels. We discussed the familiar classification methods—logistics regression, Naive Bayes, decision trees, and SVM—to give you an understanding of how these algorithms work. Oracle Advanced Analytics has a rich library of all these algorithms packaged. All you have to do is pick the best algorithm of your choice and implement it for your data set. Selection of algorithms is based on various criteria, and I touched some of the criteria that are useful to decide which classification algorithm to be used. The customer churn case study was to give you an understanding of how to approach a classification problem and integrate the results into your business operation database with ease. Based on the business requirement, the process can be automated and scheduled at regular intervals to update the findings in the database. This helps the business to stay up-to-date and make timely decisions before any unfortunate event happens.

Chapter 7 is about some of the advanced topics of data mining and the methods available in Oracle Advanced Analytics to implement them in-database.

CHAPTER 7

■ ■■ ■

Advanced Topics

"Data Scientist is the sexiest job of the 21st century," quotes Harvard Business Review. We went through many machine learning algorithms, data science techniques, and use cases, which are necessary to be a data scientist. However, this is just the beginning. Optimization techniques, feature engineering, hyperparameter optimization, and many other subjects are required to become a good data scientist. In the previous chapters, I discussed the building blocks of machine learning, data science, and ways through which we can automate certain data science tasks using Oracle Advanced Analytics. If you have read and practiced the examples and the case studies, you are all set to start your data science journey using Oracle Advanced Analytics. In this chapter, I will quickly touch on some more practical topics that are widely employed in industries and are packaged with Oracle Advanced Analytics for quick implementation of data science and analytics projects.

In this chapter, I discuss the concepts and examples for the following topics:

- Overview of neural networks

- Neural networks using Oracle Advanced Analytics

- Overview of anomaly detection

- Anomaly detection using Oracle Advanced Analytics

- Overview of predictive queries

- Predictive queries using Oracle Advanced Analytics

- Overview of the product recommendation engine

- Product recommendation engine using Oracle Advanced Analytics

- Overview of random forest

- Random forest using Oracle Advanced Analytics

© Sibanjan Das 2016
S. Das, *Data Science Using Oracle Data Miner and Oracle R Enterprise*,
DOI 10.1007/978-1-4842-2614-8_7

Overview of Neural Networks

The basic idea behind a neural network is to create a computing system that is modeled after the working of our biological brain. A neural network can learn by itself, recognize patterns out of the raw data, and make decisions: all on its own just like human brains. It is widely used in image processing for face recognition modules, in the financial industry to identify transactional frauds, and in manufacturing for automated quality control such as decisions to accept or reject a batch of steel. The applications of neural networks are immense. It forms the brain of robots and is the core component of deep learning techniques. It is estimated that deep learning would solve many mysteries that mankind has and automate huge manual tasks that were not possible ever before. In this section, I will try to touch on some basic concepts of the neural network in an illustrative way and you can write your very first neural network using Oracle Advanced Analytics.

A typical neural network is composed of many individual processing units called neurons and is organized in layers. Based on how we design a neural network, the number of neurons can range from one to a million and even more. The primary task of each neuron is to receive input, process it using some functions, and deliver the output. A neuron does the processing based on the assigned task. For example, an input neuron can only accept raw data and transmit this to other connected neurons. For the ease of classification of the neurons, they are arranged in layers. We group the layers in a neural network into three kinds: input, hidden, and output layers. Each neuron is connected to other neurons in the same layer and/or layers at either side. Figure 7-1 shows an example of a fully formed neural network with one input layer (two neurons), one hidden layer (three neurons), and one output layer (one neuron).

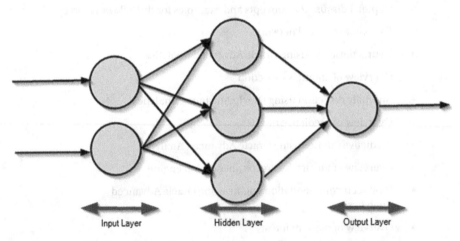

Figure 7-1. *A fully formed neural network*

> *Input Layers*: These consist of neurons that accept the input values. The output from these neurons is same as the input predictors. As a rule of thumb, the input layer has the same number of neurons as the number of predictors.

Output Layer: This is the final layer of a neural network that returns the result back to the user environment. Based on the design of a neural network, it also signals the previous layers on how they have performed in learning the information and accordingly improved their functions.

Hidden Layer(s): Hidden layers are in between input and output layers. Typically, the number of hidden layers range from one to many. It is the central computation layer that has the functions that map the input to the output of a node.

Function of a Hidden Layer Neuron

A single neuron in a hidden layer takes several inputs and produces an output as shown in Figure 7-2. Typically, they compute a weighted sum of inputs and apply certain functions to it. These functions are called "activation" or "transfer" functions. The transfer function transforms the result of the summation output to a working output using mathematical functions. Mostly, the transfer functions are differential functions to enable continuous error correction and compute the local gradient. There are many such functions such as sigmoid and tanh that can be used as a transfer function. You can find the comprehensive list of the activation functions supported by Oracle Advanced Analytics from the ORE User manual available on the OTN web site. The sigmoid function is one of the most common transfer functions used in the construction of neural networks with binary targets, as it is a strictly increasing function and its derivative is always positive.

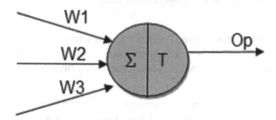

Figure 7-2. A single neuron in the hidden layer

As seen in Figure 7-2, the connections/links between neurons are associated with a weight, which is used in the calculation of the weighted sum of input in a neuron. Weights are real numbers expressing the importance of respective inputs to the output. The higher the weight of a neuron, the more influence it has on another neuron. The weights are used to control a neural network. If the neural network generates the desired output, there is no need to adjust the weights. However, if the network generates an undesirable result, or the variance between the expected and actual is high, then the weights are adjusted to improve the subsequent results.

The neural network shown in Figure 7-1 is a feed-forward neural network. It is a multilayer of neurons connected with each other that takes an input, traverses through the hidden layer, and finally reaches the output layer. There is no feedback mechanism to improve the results.

Figure 7-3 illustrates a feed-forward neural network with backpropagation learning. Backpropagation learning is an abbreviated term for the backward propagation of errors. In the case of back propagation learning, the real value is compared with the output of the neural network, and the efficiency of the weights are checked. The result is propagated back to the previous layers, and the error between the actual and expected are calculated for each node. Based on the learning, the weights are adjusted, and this process repeats until it reaches the input layer, where there is no error.

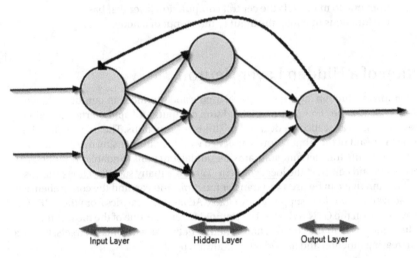

Figure 7-3. *Feed-forward neural network with backpropagation learning*

The neural network function in the Oracle Advanced Analytics package is a multilayer feed-forward neural network. Currently, neural networks are supported only in Oracle R Enterprise. It supports any number of hidden layers and an arbitrary number of neurons per layer.

■ **Note**   The ore.neural function does not preprocess the input data. Appropriate data normalization and scaling are strongly recommended.

Parameters for Neural Network

Before I move on to coding a neural network using Oracle Advanced Analytics, you need to understand some of its parameters that are essential to setting it up (see Table 7-1).

Table 7-1. *Parameters for Neural Networks*

Parameter	Description
weight	To assign the initial vector of weights; when NULL, a random starting point is generated.
hiddenSizes	To indicate the number of hidden layers and number of neurons in each hidden layer; for example, hiddenSizes = c(3, 5) specifies a neural network with two hidden layers: the first hidden layer will have 3 neurons, and the second one will have 5.
activations	This argument specifies activation functions for the hidden and the output neural network layers; it supports a single activation function per layer. If the activations argument is NULL, then the activation function for each hidden layer is bipolar sigmoid; and for the output, it is linear.
lowerBound	Lower bound for the range of values to be used by the neural network for weight initialization; it is not required if weights are supplied.
upperBound	Upper bound for the range of values to be used by the neural network for weight initialization; it is not required if weights are supplied.
supplemental.cols	Additional columns to include in the prediction result from the new data set.

Sometimes it is necessary to stop growing a neural network at some specified conditions to avoid overfitting or unnecessary multiple iterations after the objective is achieved. Table 7-2 lists some of these stopping criteria that can be introduced in a neural network to stop growing them early (and their corresponding parameter setting for the ore.neural function).

Table 7-2. *Stopping Criterias*

gradTolerance	To stop when errors are below a threshold
maxIterations	To stop when maximum number of iterations has been reached
objMinProgress	When weights and bias hardly changes, and there is a minimal relative change in the objective function value

■ **Note** While training a neural network, you may start with a single hidden layer with a small number of hidden neurons. Compute the confusion matrix and then gradually increase the number of layers and neurons until no further error reduction can be observed in the validation data set.

Neural Network Using Oracle Advanced Analytics

Neural Network is only available in the Oracle R Enterprise platform. However, you can call this procedure from either PLSQL or Oracle Data Miner workflow using the ORE and SQL integration mechanism that I discussed in earlier chapters.

> **STEP 1:** Set a seed value and load the ORE library. Setting a seed is necessary when we are using any algorithm that involves random numbers to optimize the results. You can use any seed, but it should be consistent across the task. If you do not fix the seed in creating this neural network, you might see a variation in results due to random weight initialization each time you execute the code.

```
set.seed(999)
library(ORE)
```

> **STEP 2:** Use ore.connect to connect to the database.

```
ore.connect(user = "dmuser", sid = "ORCL", host = "localhost", password =
"sibanjan123",port = 1521)
```

> **STEP 3:** Read the credit data training CSV file using the read. csv function. This converts the data into a data frame in R.

```
trainset <- read.csv("C:/Users/Admin/Dropbox/analytics_book/chapter-8/
working/creditset_train.csv")
```

> **STEP 4:** Use the ore.push function to translate the R object to corresponding ore.frame object; ore.frame maps native R objects to Oracle R specific database objects.

```
trainData <- ore.push(trainset)
```

> **STEP 5:** Invoke ore.neural with the desired parameters to create a neural network. We first try to build a neural network with one hidden layer and five neurons. As this is a binary classification problem, we use the logistics sigmoid and linear functions as transfer functions for the hidden layer and output layer, respectively.

```
fit <- ore.neural('default10yr ~ LTI + age', data = trainData,hiddenSizes =
c(5L), ,activations = c("sigmoid", "linear"))
```

> **STEP 6:** Read the credit data test CSV file into the R using the read.csv function.

```
testset <- read.csv("C:/Users/Admin/Dropbox/analytics_book/chapter-8/
working/creditset_test.csv")
```

> **STEP 7:** Use the ore.push function to translate the R object to a corresponding ore.frame object. Assign clientid as rownames for testData ore.frame. This tags a name to each row in the data set, which makes operations on rows easier.

```
testData <- ore.push(testset)
rownames(testData) <- testData$clientid
```

> **STEP 8:** Use the ore.predict function to predict the defaults for the test data.

```
ans <- ore.predict(fit, newdata = testData, supplemental.cols =
c("clientid","default10yr"))
```

> **STEP 9:** Pull out the results from the database using the ore.pull function.

```
res <- ore.pull(ans)
```

> **STEP 10:** View the prediction results of the test data on the R editor using the following code snippet.

```
res$pred_default10yr <- round(res$pred_default10yr)
rownames(res) <- res$clientid
res
```

Figure 7-4 is a snapshot of the results from the neural network.

```
> res
    clientid default10yr pred_default10yr
307       307           1                   1
308       308           0                   0
309       309           0                   1
310       310           0                   0
311       311           0                   0
312       312           0                   0
313       313           0                   1
314       314           0                   0
315       315           0                   0
316       316           0                   0
317       317           0                   0
318       318           0                   0
319       319           0                   0
320       320           0                   0
321       321           0                   0
322       322           0                   0
323       323           0                   0
324       324           0                   0
325       325           0                   0
326       326           1                   0
327       327           0                   0
328       328           1                   0
329       329           0                   0
330       330           0                   0
331       331           0                   1
332       332           0                   0
333       333           0                   0
334       334           0                   0
335       335           0                   0
336       336           0                   0
337       337           0                   0
338       338           0                   0
339       339           0                   0
340       340           0                   0
341       341           0                   0
342       342           0                   0
```

Figure 7-4. Neural network results

> **STEP 11:** Create the confusion matrix for validating the
> prediction results (see Figure 7-5). You can see from the
> confusion matrix that your first neural network has done a
> pretty good job in identifying the loan defaulters. You can
> further improve your results by trying out different number of
> hidden layers, neurons, and activation functions.

```
confusion.matrix <- with(res, table(default10yr, pred_default10yr))
confusion.matrix
```

```
> confusion.matrix
           pred_default10yr
default10yr  0  1
         0 34  3
         1  2  3
```

***Figure 7-5.** Confusion matrix*

Another possible use case for neural networks that I didn't cite in this section is anomaly detection. Anomalies are practicable in every business and occur in one or another form. For example, the anomaly detection can be a financial fraud, a network breach, or a machine failure. Other than neural networks, the Oracle Advanced Analytics platform is equipped with some state of the art algorithms for detecting anomalies from the data. In the next section, I will discuss anomalies and touch on one-class SVM, which is widely used for detecting anomalies.

Overview of Anomaly Detection

I always take a short route from home to the office and back every day. It is not well maintained but serves my purpose to reach the office on time, within 15 minutes. One fine morning, I see some large pits in the road and take a diversion from my regular route. This is an anomaly. I detected it early before driving into the holes and chose a different safe route for my trip. Had I not seen the pits or was negligent enough not to take care with it, I would have faced some serious consequences.

An anomaly is anything that occurs unexpected and is a rare event. It is a deviation from the standard pattern and does not confirm to the normal behavior of the data. Let's say we work in a steel manufacturing industry and we see the quality of the steel suddenly drops down below the permissible limits. This is an anomaly; if not detected and resolved soon, it will cost the organization millions. So to detect an anomaly at an early stage of its occurrence is very crucial.

Anomaly detection systems can be either rule-based or machine-learning-based detection systems. Rule-based systems are usually designed by defining rules that describe an anomaly and assigning thresholds and limits. Anything above the threshold or below it is classified as an anomaly. Rule-based systems are designed from the experience of the industry experts and can be thought of as systems that are designed to detect "known anomalies." They are known anomalies because we recognize what is usual and what is not usual. However, if there is something that is not known and has never been discovered so far, then creating rules for detecting anomalies is difficult. In such cases, machine-learning-based detection systems come in handy. They are a bit complex but can deal with many uncertain situations. Mostly, they can deduce the patterns that are unusual and alert the users. Many times there can be false alarms, but there are ways in machine learning where we can train the machine to reduce the false alarms.

Oracle Advanced Analytics is equipped with a one-class SVM for anomaly detections. We can also use other algorithms such as neural networks and clustering methods for this purpose.

One-Class SVM

One-class SVM is a particular case of SVM for anomaly detections. SVMs are primarily used for binary or multiclass classification data mining, which I discussed in Chapter 6. However, what if the data has only one class and the goal is to ascertain whether the new data is similar to the training data set or not? This is a typical case in anomaly detection, where we know just one class—that is, the current normal behavior—and want to find any deviations from this standard pattern, outliers, or any previously unseen class.

A one-class SVM develops a profile that describes a typical case in the training data. Deviation from the profile is flagged as an anomaly.

The type parameter for ore.odmSVM function has to be set to "anomaly.detection" while creating the SVM model for anomaly detection. The rest of the parameters are the same as those discussed for SVM in Chapter 6.

Anomaly Detection Using Oracle Advanced Analytics

What follows is a demonstration of detecting anomaly using a one-class SVM. I will use the default mtcars data set available in R. The mtcars data set comprises 10 aspects of automobile design, fuel consumption, and performance for 32 automobiles.

STEP 1: Load the ORE library and connect to the Oracle Database.

```
library(ORE)
ore.connect(user = "dmuser", sid = "ORCL", host = "localhost", password =
"sibanjan123",port = 1521)
```

STEP 2: Do the basic transformations for factor variables.

```
mtcar <- mtcars
mtcar$gear <- as.factor(mtcar$gear)
mtcar$carb <- as.factor(mtcar$carb)
mtcar$cyl <- as.factor(mtcar$cyl)
mtcar$vs  <- as.factor(mtcar$vs)
mtcar$am  <- as.factor(mtcar$am)
mtcar$ID  <- 1:nrow(mtcar)
```

STEP 3: Push the mtcar data set to the Oracle database.

```
MTCARS <- ore.push(mtcar)
```

STEP 4: Create the one-class SVM model for anomaly detection by invoking ore.odmSVM with "anomaly.detection" as a mandatory parameter. The kernel function is chosen to be linear, as the data is linearly separable. If the data is not linearly separable, use nonlinear kernels such as Gaussian.

```
svm.model <- ore.odmSVM( ~ .-ID, MTCARS,"anomaly.detection",kernel.
function="linear")
```

STEP 5: Use ore.predict to classify typical and anomalous data points. The prediction column marks anomalous records with zero (0) and typical/normal records with one (1).

```
svm.result <- ore.predict (svm.model, MTCARS, "ID")
```

STEP 6: Validate the number of anomalous and typical records in the data set. We can see three records marked as anomalous (see Figure 7-6).

```
table(svm.result$PREDICTION)
```

```
> table(svm.result$PREDICTION)

0  1
3 29
```

Figure 7-6. *Summary of the anomaly prediction*

STEP 7: Pull out the results from the Oracle database.

```
mtcar_pred <- ore.pull(svm.result)
```

STEP 8: Merge the training and prediction data set to have a holistic view of the anomalous records (see Figure 7-7).

```
results <- merge(mtcar,mtcar_pred,by="ID")
```

results

```
> results
   ID  mpg cyl  disp  hp drat    wt  qsec vs am gear carb        '1'        '0' PREDICTION
1   1 21.0   6 160.0 110 3.90 2.620 16.46  0  1    4    4 0.5056674 0.4943326          1
2   2 21.0   6 160.0 110 3.90 2.875 17.02  0  1    4    4 0.5152863 0.4847137          1
3   3 22.8   4 108.0  93 3.85 2.320 18.61  1  1    4    1 0.5102964 0.4897036          1
4   4 21.4   6 258.0 110 3.08 3.215 19.44  1  0    3    1 0.5234046 0.4765954          1
5   5 18.7   8 360.0 175 3.15 3.440 17.02  0  0    3    2 0.5344725 0.4655275          1
6   6 18.1   6 225.0 105 2.76 3.460 20.22  1  0    3    1 0.5000297 0.4999703          1
7   7 14.3   8 360.0 245 3.21 3.570 15.84  0  0    3    4 0.5057617 0.4942383          1
8   8 24.4   4 146.7  62 3.69 3.190 20.00  1  0    4    2 0.5003826 0.4996174          1
9   9 22.8   4 140.8  95 3.92 3.150 22.90  1  0    4    2 0.5125569 0.4874431          1
10 10 19.2   6 167.6 123 3.92 3.440 18.30  1  0    4    4 0.5012371 0.4987629          1
11 11 17.8   6 167.6 123 3.92 3.440 18.90  1  0    4    4 0.4999888 0.5000112          0
12 12 16.4   8 275.8 180 3.07 4.070 17.40  0  0    3    3 0.5029513 0.4970487          1
13 13 17.3   8 275.8 180 3.07 3.730 17.60  0  0    3    3 0.5045981 0.4954019          1
14 14 15.2   8 275.8 180 3.07 3.780 18.00  0  0    3    3 0.5000226 0.4999774          1
15 15 10.4   8 472.0 205 2.93 5.250 17.98  0  0    3    4 0.5000612 0.4999388          1
16 16 10.4   8 460.0 215 3.00 5.424 17.82  0  0    3    4 0.5034760 0.4965240          1
17 17 14.7   8 440.0 230 3.23 5.345 17.42  0  0    3    4 0.5291758 0.4708242          1
18 18 32.4   4  78.7  66 4.08 2.200 19.47  1  1    4    1 0.5124063 0.4875937          1
19 19 30.4   4  75.7  52 4.93 1.615 18.52  1  1    4    2 0.4955863 0.5044137          0
20 20 33.9   4  71.1  65 4.22 1.835 19.90  1  1    4    1 0.5057599 0.4942401          1
21 21 21.5   4 120.1  97 3.70 2.465 20.01  1  0    3    1 0.5044280 0.4955720          1
22 22 15.5   8 318.0 150 2.76 3.520 16.87  0  0    3    2 0.5000473 0.4999527          1
23 23 15.2   8 304.0 150 3.15 3.435 17.30  0  0    3    2 0.5183815 0.4816185          1
24 24 13.3   8 350.0 245 3.73 3.840 15.41  0  0    3    4 0.5158205 0.4841795          1
25 25 19.2   8 400.0 175 3.08 3.845 17.05  0  0    3    2 0.5374437 0.4625563          1
26 26 27.3   4  79.0  66 4.08 1.935 18.90  1  1    4    1 0.5000008 0.4999992          1
27 27 26.0   4 120.3  91 4.43 2.140 16.70  0  1    5    2 0.5180578 0.4819422          1
28 28 30.4   4  95.1 113 3.77 1.513 16.90  1  1    5    2 0.5000287 0.4999713          1
29 29 15.8   8 351.0 264 4.22 3.170 14.50  0  1    5    4 0.5353505 0.4646495          1
30 30 19.7   6 145.0 175 3.62 2.770 15.50  0  1    5    6 0.4964244 0.5035756          0
31 31 15.0   8 301.0 335 3.54 3.570 14.60  0  1    5    8 0.5000297 0.4999703          1
32 32 21.4   4 121.0 109 4.11 2.780 18.60  1  1    4    2 0.5304042 0.4695958          1
```

Figure 7-7. Results displaying the anomalous automobiles

The results clearly show the automobiles that were marked anomalous. We can also use the anomaly detection query option available in the Predictive Queries section of the Oracle SQL Developer GUI. We need not know the machine learning techniques to build an anomaly detection model using this option. This method is one of the useful features that the Oracle Advanced Analytics platform offers and that I discuss in the next section.

Overview of Predictive Analytics in Oracle Advanced Analytics

Predictive queries and packages are used to create on the fly predictive models for data in the Oracle Database. For using the predictive analytics packages and functionalities, you need not learn the intricacies and mathematical complexity of machine learning algorithms. Instead, you can directly supply your data sets to the predictive analytics engine and view the results. The predictive analytics engine understands the data, trains the models, tests it, and then generates the results. However, these models are not preserved after the operation is complete and the result is returned to the user. There are many aspects of data science that predictive analytics doesn't address, such as optimization and regularization of algorithms, but we can say it is just the beginning of the revolution for automated data science. A user or an organization who has no knowledge of data science can use these APIs for implementing predictive analytics solutions. If you are a data scientist or a machine learning expert, you can use the predictive analytics packages and queries to quickly build a machine learning model. It can come in handy when you want to do a quick proof of the concept for winning the confidence of stakeholders before the actual start of a project.

In the following sections, I will demonstrate the predictive analytics package and the related functions that Oracle Advanced Analytics offers.

Predictive Analytics Using SQL Developer GUI

The Predictive Queries choice is available in the component panel of the data miner option of SQL Developer. Table 7-3 displays the four functionalities that can be performed on the input data sets for predictive analytics.

Table 7-3. *Predictive Query Techniques*

Predictive Query Technique	Description
Prediction Query	Creates regression/classification models based on the type of target attribute and input data set. If the target is a numeric value, it performs regression; and if it is of character data type, it performs classification.
Anomaly Detection Query	Builds an anomaly detection model on an input data set and returns the results indicating the anomalous records.
Feature Extraction Query	Automatically identifies and extracts key features from the input data set.
Clustering Query	Forms clusters for the input data set automatically and returns the data with an assigned cluster for each record.

I will use the customer churn data set used in Chapter 6 to demonstrate the prediction query functionality of the predictive queries. You can try out other predictive queries options similarly.

STEP 1: Define the DEMO_CUST_CHURN_TAB data source by dragging the data source node from the data section of the components panel and selecting the table DEMO_CUST_CHURN_TAB. Then drag Prediction Query node from the Predictive Queries panel to the workflow editor and connect it to the DEMO_CUST_CHURN_TAB data source as shown in Figure 7-8.

DEMO_CUST_CHURN_TAB   Prediction Query

Figure 7-8. *Data source node connected to Prediction Query*

STEP 2: Right-click on the Prediction Query Node and select the edit option. As shown in Figure 7-9, select the Case ID as CUSTOMERID and add CHURN as the target attribute.

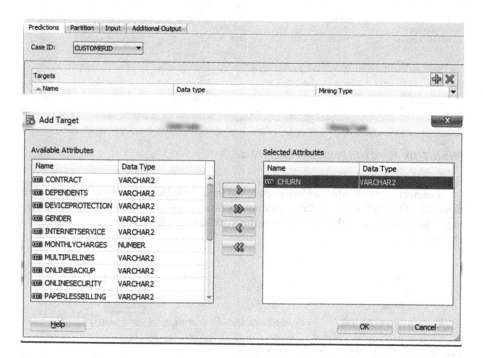

Figure 7-9. *Edit options in prediction query*

Select OK to exit the Add Target screen. The Edit Prediction Query node should look like Figure 7-10. You can see the prediction output section in the screen automatically populated with Prediction, Prediction Details, and Prediction Probability. These three attributes would provide you with the prediction results, the rule used for prediction, and the probability associated with the prediction, respectively.

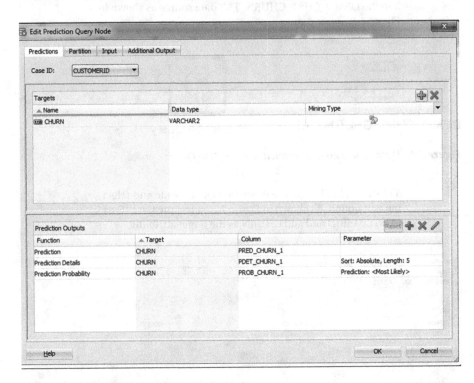

Figure 7-10. *Final Edit Prediction Query Node*

Connect a Create Table or View node to the Prediction Query node and assign a table name DEMO_CUST_CHURN_PRED to store the results. You can follow the steps outlined in previous chapters for this task. Execute the workflow. The final workflow looks like Figure 7-11.

DEMO_CUST_CHURN_TAB Prediction Query DEMO_CUST_CHURN_PRED

Figure 7-11. *Final Prediction Query workflow*

Once completed, you can view the prediction results from the DEMO_CUST_
CHURN_PRED table (Figure 7-12).

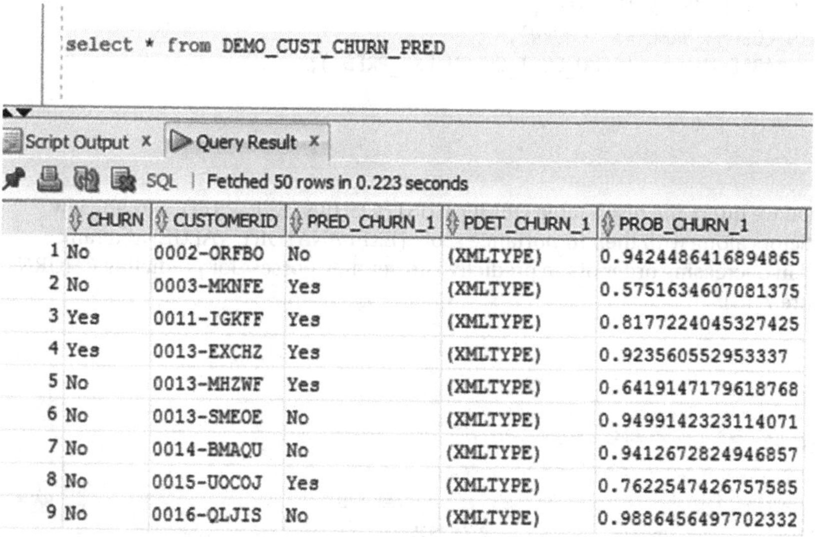

```
select * from DEMO_CUST_CHURN_PRED
```

Script Output × ▷ Query Result ×

SQL | Fetched 50 rows in 0.223 seconds

	CHURN	CUSTOMERID	PRED_CHURN_1	PDET_CHURN_1	PROB_CHURN_1
1	No	0002-ORFBO	No	(XMLTYPE)	0.9424486416894865
2	No	0003-MKNFE	Yes	(XMLTYPE)	0.5751634607081375
3	Yes	0011-IGKFF	Yes	(XMLTYPE)	0.8177224045327425
4	Yes	0013-EXCHZ	Yes	(XMLTYPE)	0.923560552953337
5	No	0013-MHZWF	Yes	(XMLTYPE)	0.6419147179618768
6	No	0013-SMEOE	No	(XMLTYPE)	0.9499142323114071
7	No	0014-BMAQU	No	(XMLTYPE)	0.9412672824946857
8	No	0015-UOCOJ	Yes	(XMLTYPE)	0.7622547426757585
9	No	0016-QLJIS	No	(XMLTYPE)	0.9886456497702332

Figure 7-12. Results

You can also create an on the fly predictive analytics model using SQL and PLSQL
API. These are useful when you want to integrate the models with any data visualization
tools or any front-end applications. Almost all front-end software and data visualization
tools have PLSQL support, and this makes it easy to leverage the value of on the fly, in-
database predictive models provided by the Oracle Advanced Analytics platform. The
DBMS_PREDICTIVE_ANALYTICS package provides the methods necessary to perform
these operations.

DBMS_PREDICTIVE_ANALYTICS

There are three main operations that can be performed using the DBMS_PREDICTIVE_
ANALYTICS predictive analytics package: EXPLAIN, PROFILE, and PREDICT.

EXPLAIN

The EXPLAIN function is used to explain attributes that are important for a model. It
returns the ranks and importance of the attributes in relative order of their importance in
predicting the target attribute. Behind the scenes, it implements the attribute importance
functionality that I discussed in Chapter 6.

```
BEGIN
  DBMS_PREDICTIVE_ANALYTICS.EXPLAIN(
  DATA_TABLE_NAME => 'DEMO_CUST_CHURN_TAB',
  DATA_SCHEMA_NAME => 'DMUSER',
  EXPLAIN_COLUMN_NAME => 'CHURN',
  RESULT_TABLE_NAME =>'DEMO_CUST_CHURN_EXP_PRED');

END;
/
```

When we query the result table DEMO_CUST_CHURN_EXP_PRED, we can view the predictors along with their importance score (EXPLANATORY_VALUE) and rank arranged in decreasing order of the predictive power they possess for predicting CHURN (see Figure 7-13).

```
select * from DEMO_CUST_CHURN_EXP_PRED
```

Query Result × Script Output × Query Result 1 ×

SQL | All Rows Fetched: 20 in 0.019 seconds

	ATTRIBUTE_NAME	ATTRIBUTE_SUBNAME	EXPLANATORY_VALUE	RANK
1	CONTRACT	(null)	0.169985446520973421208228131005654601826 8	1
2	TENURE	(null)	0.121035771796872437610040287486531579067 4	2
3	ONLINESECURITY	(null)	0.108060217438743831992131666463873145995 7	3
4	TECHSUPPORT	(null)	0.102825918902879498438492904786787527844 9	4
5	INTERNETSERVICE	(null)	0.090357146900932220738169747525261417361 2	5
6	ONLINEBACKUP	(null)	0.075174004107229213025633169744827090379 8	6
7	DEVICEPROTECTION	(null)	0.072467586197264552519646627877735203090 8	7
8	PAYMENTMETHOD	(null)	0.071493615765924985723611868732274722978 2	8
9	MONTHLYCHARGES	(null)	0.067598516389438301604788612771046852388 2	9
10	TOTALCHARGES	(null)	0.061201734890535722216659499681245508665 9	10
11	STREAMINGTV	(null)	0.051158666116074503257110209610322281339	11
12	STREAMINGMOVIES	(null)	0.051158666116074503257110209610322281339	11
13	PAPERLESSBILLING	(null)	0.030873730484055927469947481011433701777 3	12
14	DEPENDENTS	(null)	0.022626958680430026263036691007315165650 4	13
15	PARTNER	(null)	0.018045095972440889565069276107121854204 8	14
16	SENIORCITIZEN	(null)	0.015808870932381681303247156418385746262 5	15
17	MULTIPLELINES	(null)	0	16

Figure 7-13. Explain method results

PREDICT

The PREDICT method produces predictions for unknown targets. Based on the values of the target attribute, it decides whether to opt for a regression algorithm or classification algorithms. If the target is a numeric attribute, it creates a regression model; and if it is a categorical attribute, it creates a classification model. The input data to the PREDICT method should contain records for both the training as well as new data sets. For the new data set, the target attribute should be NULL or set to NULL. Any case in which the target is unknown, that is, when the target value is null, will not be considered during model training.

```
DECLARE
  v_accuracy NUMBER(30,10);
BEGIN
  DBMS_PREDICTIVE_ANALYTICS.PREDICT(
  ACCURACY => v_accuracy,
  DATA_TABLE_NAME => 'DEMO_CUST_CHURN_TAB',
  DATA_SCHEMA_NAME => 'DMUSER',
  CASE_ID_COLUMN_NAME => 'CUSTOMERID',
  TARGET_COLUMN_NAME => 'CHURN',
  RESULT_TABLE_NAME => 'DEMO_CUST_CHURN_ANALYTICS_PRED');

  dbms_output.put_line('*** Accuracy ***');
  dbms_output.put_line(' Accuracy '||v_accuracy);
END;
/
```

After we execute the code, we can see the accuracy measure, which in the AUC metrics is 0.4993768397. This is an average score, but there is no way to improve this accuracy using this API (see Figure 7-14). It is a demerit of the predictive analytics package, but it is there for automated machine learning.

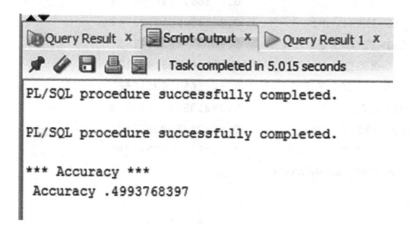

Figure 7-14. *Accuracy of the PREDICT method classifier*

When you query the result table DEMO_CUST_CHURN_ANALYTICS_PRED, you can find the predictions for each CUSTOMERID along with their prediction probability (see Figure 7-15).

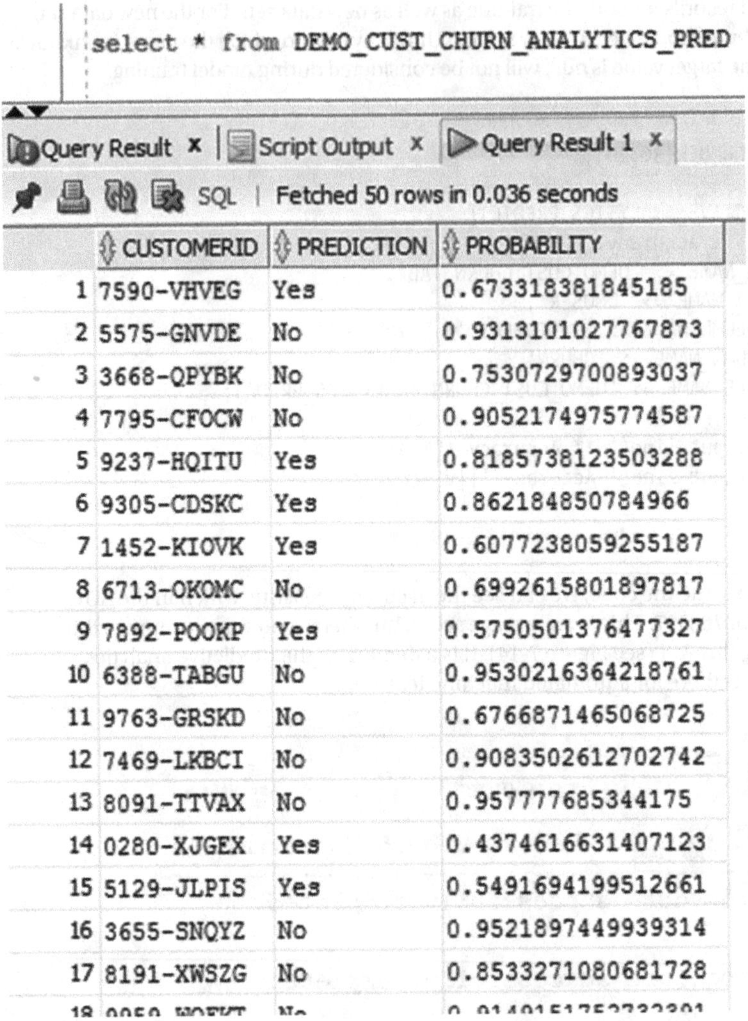

```
select * from DEMO_CUST_CHURN_ANALYTICS_PRED
```

▲▼

🔲 Query Result ✕ | 🔲 Script Output ✕ | ▷ Query Result 1 ✕

📌 🖨 🔍 📋 SQL | Fetched 50 rows in 0.036 seconds

	◊ CUSTOMERID	◊ PREDICTION	◊ PROBABILITY
1	7590-VHVEG	Yes	0.673318381845185
2	5575-GNVDE	No	0.9313101027767873
3	3668-QPYBK	No	0.7530729700893037
4	7795-CFOCW	No	0.9052174975774587
5	9237-HQITU	Yes	0.8185738123503288
6	9305-CDSKC	Yes	0.862184850784966
7	1452-KIOVK	Yes	0.6077238059255187
8	6713-OKOMC	No	0.6992615801897817
9	7892-POOKP	Yes	0.5750501376477327
10	6388-TABGU	No	0.9530216364218761
11	9763-GRSKD	No	0.6766871465068725
12	7469-LKBCI	No	0.9083502612702742
13	8091-TTVAX	No	0.957777685344175
14	0280-XJGEX	Yes	0.4374616631407123
15	5129-JLPIS	Yes	0.5491694199512661
16	3655-SNQYZ	No	0.9521897449939314
17	8191-XWSZG	No	0.8533271080681728
18	0050 WOFUT	No	0.01401517007200201

Figure 7-15. *PREDICT method results*

PROFILE

The PROFILE method is to create profiles or rules that provide insights on the reason for a particular prediction. Using PROFILE, we can see the patterns that define the outcome. For example, we can create customer profiles based on their churn status (yes/no). Using the PROFILE method can help us retrieve rules that define the attributes and behavior of customers who have churned or have stayed with us. Behind the scenes, it executes the decision tree algorithm to accomplish this task.

```
BEGIN
  dbms_predictive_analytics.profile(
  DATA_TABLE_NAME => 'DEMO_CUST_CHURN_TAB',
  data_schema_name => 'DMUSER',
  target_column_name => 'CHURN',
  result_table_name =>'DEMO_CUST_CHURN_PROF_PRED');

END;
```

When we query the results table DEMO_CUST_CHURN_PROF_PRED, we can view the various rules that describe each score (or prediction). The result/description is of XML type. Click on the pencil icon next to the Description column in Query Results to view the XML (see Figure 7-16).

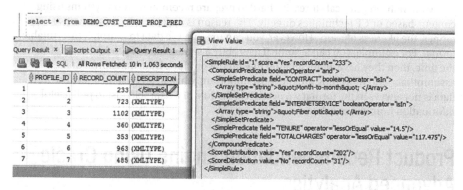

Figure 7-16. *PROFILE method results*

Overview of Product Recommendation Engine

A product recommendation system provides suggestions for products to a user. The recommendations relate to various decision-making processes such as what items to buy or what service to avail. A recommendation system provides useful and practical suggestions for the particular type of product that is of benefit to a user.

The goal of this system is to provide right and relevant product recommendations for customers so that the probability of sales conversion is high.

Product recommendation systems are widely used in e-commerce applications. However, they can also be leveraged for physical stores. There are various types of recommendation systems. Following are two widely used recommendation system designs in the e-commerce space:

1. ***Content based***: Examine properties of the items to recommend items that are similar in content to items the user has already liked in the past or matched to the attributes of the user.

2. ***Collaborative filtering (CF)***: A user is recommended items based on the past rating of all users collectively. CF can be of two types:

    a. ***User-based CF***: Given a user U, find the set of other users D whose ratings are similar to the ratings of the selected user U.

    b. ***Item-based CF***: Build an item-item matrix determining relationships between pairs of items.

The design of a recommender system is not limited to these types. We can also use association rules and the Naïve Bayes algorithm. We can also prepare a hybrid recommendation using multiple techniques to further narrow down and improve the quality of recommendations.

Generally, in physical stores, it's hard to prepare recommendation systems using content-based or CF techniques directly. The reason is the absence of data such as ratings, the likeness of users. However, you can prepare the data to morph it to a suitable format. For example, you can assume if users buy a similar kind of product most of the time, then they have an inclination toward the product. Data transformations, assumptions, and morphing are the most important job of a data scientist. Real-world data is not always in the format we need. It is the role of a data scientist to pull useful information from the raw data.

Product Recommendation Engine Using Oracle Advanced Analytics

The Oracle Advanced Analytics platform has not yet shipped the recommendation algorithms such as content-based or CF algorithms. However, market basket analysis and the Naïve Bayes algorithm can also be used to design a recommendation system.

I continue with the case study discussed in Chapter 4 for market basket analysis. We can use the same data set DEMO_ASSOC_PUR_V that was prepared for that exercise.

> ***STEP 1:*** Define the DEMO_ASSOC_PUR_V data source by dragging the data source node from the data section of the components panel and selecting the table DEMO_ASSOC_ PUR_V. Then, drag the Filter Rows node from the transforms section to the workflow editor and connect it to the DEMO_ ASSOC_PUR_V data source (see Figure 7-17).

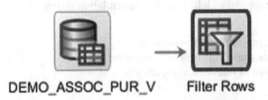

DEMO_ASSOC_PUR_V   Filter Rows

Figure 7-17. *Data source node connected to Filter Rows node*

STEP 2: Right-click on the node to edit its properties and add the filter condition as customer_id = 210342. Here, we are hard-coding the customer id to generate a recommendation for this particular customer. However, we can make it parameterized using scripts to input the customer id at the run time. This is left as an exercise for the readers to try using the techniques we discussed in Chapter 4.

STEP 3: Drag the Class Build node from the Models panel to the workflow editor. Connect it to the Filter Rows and edit its properties. Select ITEM as the target attribute, as we want to predict which item is suitable for the customer to recommend. In the Model Settings section, remove all algorithms except Naïve Bayes (see Figure 7-18).

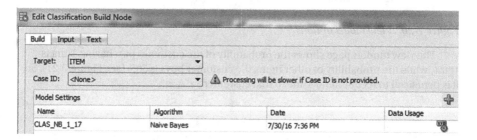

Figure 7-18. Edit Classification Build Node options

STEP 4: Execute the workflow as described in previous chapters. The final workflow should look like the Figure 7-19.

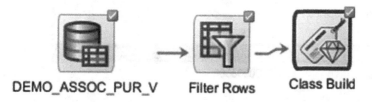

Figure 7-19. Final product recommendation workflow

STEP 5: View the model by selecting the model name from the drop-down menu of the Class Build node as shown in Figure 7-20.

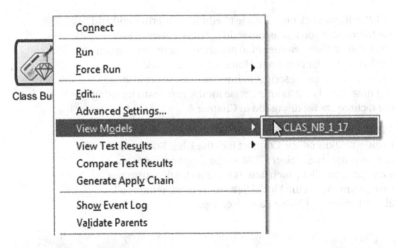

Figure 7-20. *View model options*

The view models page shows the probability of the customer purchasing Tuna is 100%, while the probability of other items the customer buys is 50% (see Figure 7-21). So, it is beneficial to recommend Tuna to that customer.

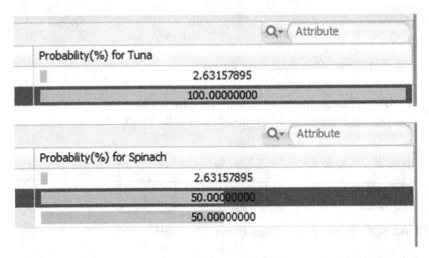

Figure 7-21. *Purchase probabilities of Tuna and Spinach*

Overview of Random Forest

Random forest is an ensemble learning method for decision trees. Ensemble models are composed of several supervised learning models that are independently trained, and the results are combined in different ways to obtain the final prediction result. The result has better predictive performance than the results of any of its constituent learning algorithms separately. Mostly there are two types of ensemble learning methods that are widely used:

1. ***Bagging***: Also known as bootstrap aggregation, it is the way to decrease the variance error of a model's result. In this method, random samples are prepared from training data set using the bootstrapping method (random sample with replacement models). Models are built on each sample using supervised learning methods. Finally, the results are combined by averaging the results or selecting the best prediction using majority voting. Majority voting is a process in which the class with the largest number of predictions in all of the classifiers becomes the prediction of the ensemble.

2. ***Boosting***: The boosting method is an iterative process in which successive models are created one after the other based on the errors of the predecessors. This helps to reduce both variance and bias in the data set. Boosting attempts to create new classifiers that are better able to predict, for which the current ensemble's performance is weak. Unlike bagging, the resampling of the training data is dependent on the performance of the earlier classifiers. Boosting uses all data to train each classifier, but instances that were misclassified by the previous classifiers are given more weight so that subsequent classifiers improve the results.

Random forest is based on the bagging method of ensemble models. Here, multiple decision trees are created with the different set of predictors. Each tree gives its prediction results. The random forest uses the voting mechanism to select the most voted prediction class as the final prediction. This produces better prediction results than a single decision tree and is more stable.

Random Forest Using Oracle Advanced Analytics

In the following example, I will illustrate how to use random forest to classify data.

STEP 1: Set a seed value and load the ORE library.

```
library(ORE)
ore.connect(user = "dmuser", sid = "ORCL", host = "localhost", password =
"sibanjan123",port = 1521)
```

STEP 2: Read the credit data training CSV file using the read. csv function.

```
trainset <- read.csv("D:/studies/neuralnet/creditset_train.csv")
testset <- read.csv("D:/studies/neuralnet/creditset_test.csv")
```

STEP 3: Use the ore.push function to translate the R object to a corresponding ore.frame object.

```
trainData <- ore.push(trainset)
testData <- ore.push(testset)
```

> **STEP 4:** Use ore.randomForest to create a random forest
> model with the parameters discussed in the previous section.

```
mod <- ore.randomForest('default10yr ~ LTI + age', data = trainData,na.
action = na.exclude)
```

> **STEP 5:** Use ore.predict to predict the values for test data.

```
ans <- ore.predict(mod, testData, type="all", supplemental.cols=c("clientid
","default10yr"))
```

> **STEP 6:** Validate the model by creating a confusion matrix
> (see Figure 7-22).

```
> confusion.matrix
           prediction
default10yr  0  1
          0 36  1
          1  0  5
```

Figure 7-22. *Confusion matrix*

```
confusion.matrix <- with(ans, table(default10yr, prediction))
```

```
confusion.matrix
```

The confusion matrix shows a precision of 83.33% in predicting loan defaulters
(values with class label as one). Though this shows a very optimistic picture of our model
in predicting defaulters, this might also be a case of overfitting.

Summary

In this chapter, I discussed neural networks, which are gaining a lot of attention and importance in the data science community due their ability to solve complex, nonlinear problems that are difficult to be performed by general machine learning algorithms. We also came across anomaly detection theory, and the corresponding Oracle Advanced Analytics packages that can be leveraged for detecting anomalies in the data. Using the customized and state-of-the-art anomaly detection packages in Oracle Advanced Analytics runs in the database and detects anomalies as and when any abnormality occurs. We also learned about ensemble methods and implemented random forest, which is widely used across the data science community for producing superior results. Until now, I have discussed the different ways we can implement and automate the data science tasks using Oracle Advanced Analytics. In Chapter 8, I discuss the different deployment and migration techniques for moving the models and workflows across environments.

CHAPTER 8

■ ■ ■

Solutions Deployment

We are often required to move the models and their associated objects to different environments in their lifetime. For example, we develop a model in a development environment, move to its test environment for user testing, and make it live in a production environment. It is not feasible to create the models from scratch in every environment, and doing so might introduce human errors at certain instances.

In this chapter, I cover the following topics:

- Techniques to export and import an Oracle Data Miner (ODM) workflow

- Explore the deploy option in SQL Developer for implementing ODM workflows

- Import and export data mining models

- Using PMML (Predictive Model Markup Language) to export and import predictive models to the Oracle Advanced Analytics platform

Export and Import an ODM Workflow

You can use the export and import workflow option in SQL Developer to export the data mining workflow and recreate it in a different environment with ease. You can also utilize this functionality to store a backup copy of your workflow as and when you do some significant enhancements on it. The development environment is a place where a lot of people do many experiments, which makes it vulnerable to database misconfiguration and prone to errors. Having a backup of the workflow at regular intervals would certainly help you preserve all the hard work and efforts you put in to developing the workflow. The export workflow option generates an XML file that can be stored and can be imported using SQL Developer.

To export a workflow, you can follow these steps:

- Right-click on the name of the workflow in the Data Miner tab. It pops up a menu having "Export Workflow" as an option as shown in Figure 8-1.

© Sibanjan Das 2016

S. Das, *Data Science Using Oracle Data Miner and Oracle R Enterprise*, DOI 10.1007/978-1-4842-2614-8_8

Figure 8-1. Export Workflow option

- Select the Export Workflow option. Enter the location and file
  name where you want the workflow to be saved. By default, the
  name of the file is the workflow name. In case you want it to
  be stored by some other name, you can change the file name
  (Figure 8-2).

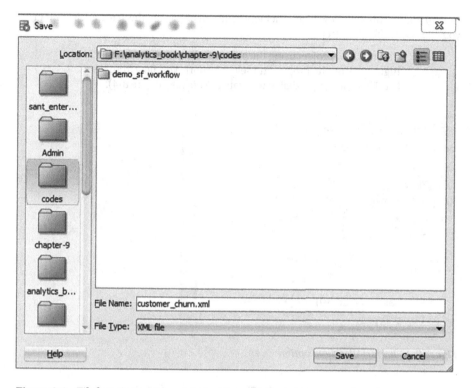

Figure 8-2. File location window to save the wokflow

- Once the workflow is migrated successfully, you can find the XML file stored in the mentioned location (see Figure 8-3).

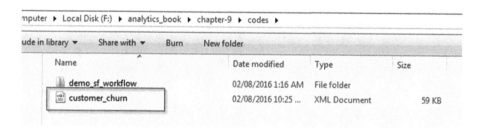

Figure 8-3. Exported XML of the workflow

To import this workflow to another environment or back to the same environment, you can follow the these steps:

- Create a new project using the process I discussed in Chapter 2. Right-click on the project and select Import Workflow from the list. This opens up a dialog to locate the file (see Figure 8-4).

Figure 8-4. Import Workflow option

- Choose the file location, the file name to be imported, and select Open as shown in Figure 8-5.

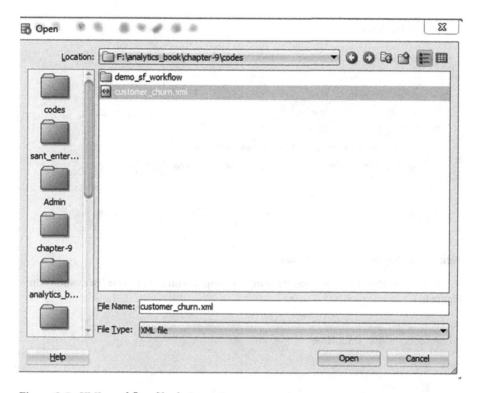

Figure 8-5. XML workflow file choice options

- The next screen confirms the details related to the workflow name, models, and output tables. These details have defaulted from the workflow XML file. Proceed with the default names or rename the workflow as per choice and click OK to continue (see Figure 8-6).

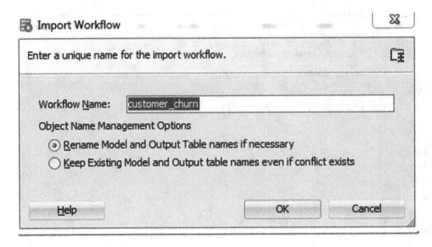

Figure 8-6. Import Workflow window

Once completed successfully, the workflow appears on the projects list as shown in Figure 8-7.

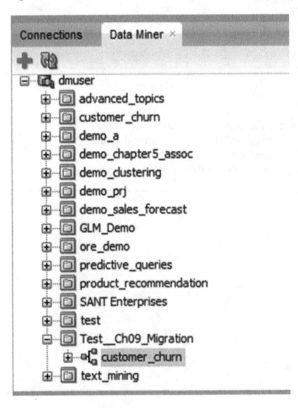

Figure 8-7. Imported customer churn workflow

Deploy Option for Data Miner Workflows in SQL Developer

The deploy option in the SQL Developer can be used to deploy the model in a production environment. We always like to automate the tasks in a production environment whenever possible. For example, we might want to schedule the models to run during the night or on off-business hours. The deploy options help to create PLSQL scripts that have all the functionalities that you have developed using the GUI option of the Oracle Advanced Analytics platform. The master script can be executed by a user or scheduled to run at intervals.

■ **Note**   The script scheduler should have the feature to execute SQL scripts or use DBMS_SCHEDULER PL/SQL API for scheduling.

To create these migration scripts, open the workflow in the SQL developer. Right-click on a node and select Deploy as shown in Figure 8-8. The deploy option shows three choices to choose. Table 8-1 lists a brief description of the options:

Table 8-1. *Deploy Options*

Deploy Options	Description
Selected node and dependent nodes	Migration scripts are created for the node that is selected for deployment along with its dependent node definitions.
Selected node, dependent nodes, and children nodes	Migration scripts are created for the selected node, nodes that are dependent on it, and also its children nodes.
Selected node and connected nodes	Migration scripts are created for the selected nodes and all the nodes that are connected to it.

We always try out different options in the workflow in our development phase. It is always advisable to clean up the workflow or create a new one with the final design before preparing the migration scripts. Selected node, dependent nodes, and children nodes are the most suitable options that I use almost every time. It migrates all the nodes that a workflow has. Also, I always edit the last node to deploy, so that we don't miss any developed functionality while migrating the codes.

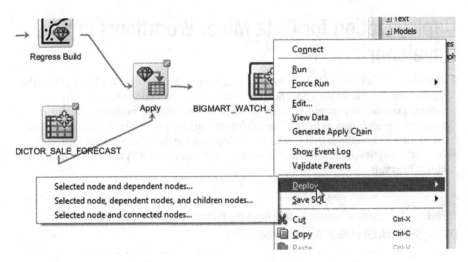

Figure 8-8. Deploy options selection menu

Once you select the "Selected node, dependent nodes, and children nodes" option, the Generate SQL Script screen pops up.

STEP 1: Choose the target database option.

The default database version in which the workflow is created is shown on the screen. It is always advisable to have the workflow development and target database version similar to minimize the chance of compatibility issues. Oracle keeps enhancing the functionalities at almost each version release, which might introduce some potential release issues. Click Next to continue to Step 2 of the SQL script generation (see Figure 8-9).

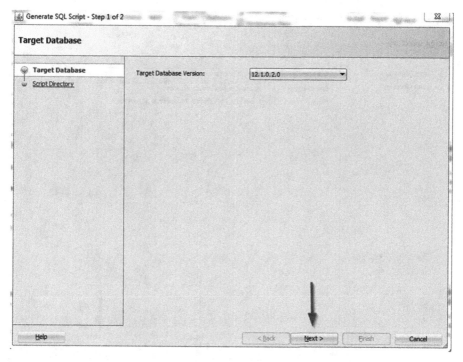

Figure 8-9. *Step to select Target Database Version*

> **STEP 2:** In this screen, you need to enter the Script Directory
> and Base Directory (see Figure 8-10). The Script directory is
> the name of the folder where all the generated scripts will be
> stored. The Base directory is the path to the Script Directory.

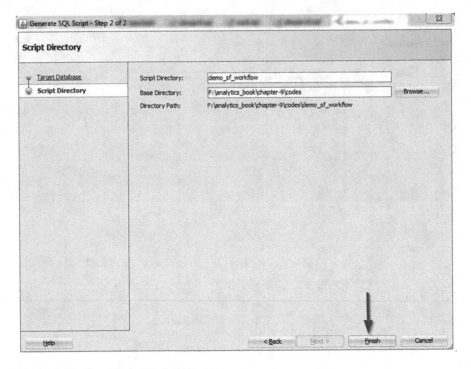

Figure 8-10. *Step to select Script Directory*

Click Finish to generate the scripts. If the scripts are created successfully, the successful message in Figure 8-11 appears on the screen.

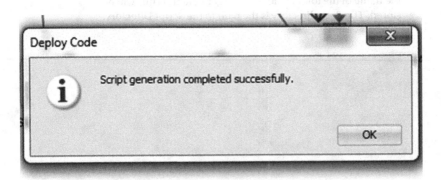

Figure 8-11. *Successful Deploy Code message*

Navigate to the script directory to check the generated scripts. You can see many scripts are created for the single workflow. The following is a brief list of the types of scripts that get created (see also Figure 8-12).

- ***<workflow>.png***: It has the image of the workflow.

- ***<node>.SQL***: An SQL script gets generated for each node of the workflow.

- ***<workflow_run>.sql***: It is the master script that is used to re-create and run the workflow. This is the script you need to execute in the target instance. This script has calls to other scripts in necessary order.

- ***<workflow_drop>.sql***: This is the cleanup script to drop the objects related to the workflow.

Name	Date modified	Type	Size
er ▸ Local Disk (F:) ▸ analytics_book ▸ chapter-9 ▸ codes ▸ demo_sf_workflow			
⊓ library ▼ Share with ▼ Burn New folder			
apply	02/08/2016 1:16 AM	SQL File	4 KB
bigmart_sale_orders	02/08/2016 1:16 AM	SQL File	3 KB
bigmart_watch_sale_forecast	02/08/2016 1:16 AM	SQL File	4 KB
demo_sf_workflow	02/08/2016 1:16 AM	PNG Image	33 KB
demo_sf_workflow_drop	02/08/2016 1:16 AM	SQL File	4 KB
demo_sf_workflow_run	02/08/2016 1:16 AM	SQL File	6 KB
filter_columns	02/08/2016 1:16 AM	SQL File	6 KB
predictor_sale_forecast	02/08/2016 1:16 AM	SQL File	4 KB
regress_build	02/08/2016 1:16 AM	SQL File	12 KB
sql_query	02/08/2016 1:16 AM	SQL File	4 KB

Figure 8-12. *Generated deployment scripts for demo_sf_Workflow*

■ **Note**   The PLSQL packages, R scripts, and shared database objects have to be migrated and compiled in the target environment prior executing the workflow scripts.

Import and Export Data Mining Models

In the preceding sections, we learned about exporting and importing the data mining workflow. What if the models are created using PLSQL and ORE ODM APIs? The export and import functionality for data mining models helps us in migrating the models between the environments with ease. It can also be used to deploy the models in another environment along with all the model and algorithm settings that we used in the development environment.

The export_model option of the DBMS_DATA_MINING package creates a dump (.dmp) file that can be used to recreate the models.

EXPORT_MODEL

The export_model procedure of the DBMS_DATA_MINING package has some parameters described in Table 8-2 to be entered for exporting the model(s).

Table 8-2. Export Model Parameters

Parameters	Description
filename	File name for the dump file.
Directory	Directory name of the location to create the dump file.
model_filter	Enter the model name as filter criteria; if you don't specify this parameter, all models in the schema are exported.
Filesize	An optional parameter; the default size is 50 MB. You can use this parameter to specify the maximum size of a file operation.
remote_link	An optional parameter to specify the remote database link.
jobname	An optional parameter that specifies the name of the export job.

The files created by Export option are physical files (files stored in the operating system). So, the DBA (Database Administrator) needs to give the DM (data mining) user access to create directories, or he/she should give read/write permission to the DM user for dumping the files into the file directory of interest as described in the following steps.

> *STEP 1:* (DBA task) The data mining user should have the permission to create any directory. The DBA can grant permission to DMUSER to create the directory using the following command:

```
grant create any directory to dmuser;
```

> *STEP 2:* From the data mining user (DMUSER), create the directories. You need to make sure the physical directory exists.

```
CREATE OR REPLACE DIRECTORY dmdir AS 'F:\sample_mig\oracle'
```

You can export a single model or multiple models by entering the model names as shown in the following codes.

Code snippet for exporting single model

```
BEGIN
DBMS_DATA_MINING.EXPORT_MODEL ('CLAS_DT_1_14.dmp','dmdir','name =''CLAS_
DT_1_14''');
END;
```

Code snippet for exporting multiple models

```
begin
DBMS_DATA_MINING.EXPORT_MODEL ( 'multi_model_out', 'dmdir','name in (''CLAS_
SVM_1_14'', ''CLAS_GLM_1_14'')');
end;
```

After the procedure is executed successfully, the dump files (files with DMP extension) and their logs show up in the specified directory (see Figure 8-13).

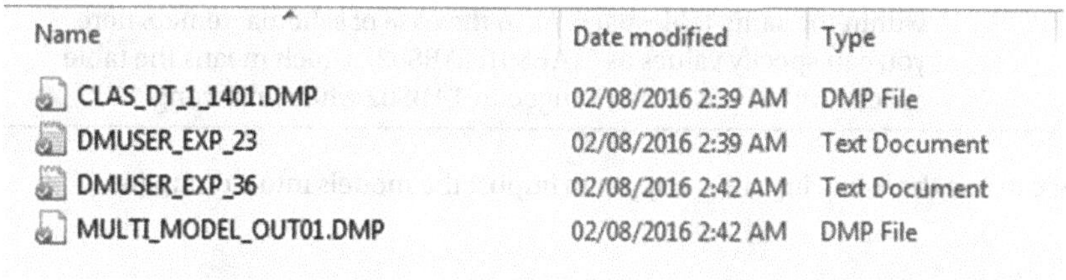

Name	Date modified	Type
CLAS_DT_1_1401.DMP	02/08/2016 2:39 AM	DMP File
DMUSER_EXP_23	02/08/2016 2:39 AM	Text Document
DMUSER_EXP_36	02/08/2016 2:42 AM	Text Document
MULTI_MODEL_OUT01.DMP	02/08/2016 2:42 AM	DMP File

Figure 8-13. *Export dump files*

IMPORT_MODEL

The dump file can be used to import the models of another schema or database using the IMPORT_MODEL procedure of the DBMS_DATA_MINING package. Table 8-3 lists the parameters that are used to control the import operations.

Table 8-3. *Import Model Parameters*

Parameters	Description
filename	Name of the dump file from which the model(s) are to be imported; the file should have been created by the export functionality.
directory	Name of the directory where the file is located.
model_filter	An optional parameter; when there are multiple models in a single dump file, you can use this option to filter models to be created in an environment.
operation	An optional parameter; there are two options to select: IMPORT and SQL_FILE. IMPORT is the default option. It is the parameter that instructs the API to import the model into the environment. The SQL_FILE option creates SQL DDL for the model creation to a text file.
remote_link	An optional parameter; it is used to specify the remote database link.
jobname	An optional parameter; you can use this option to specify the job name for the import activity of your choice. This parameter overrides the system generated name of the import job.

(continued)

Table 8-3. *(continued)*

Parameters	Description
schema_remap	An optional parameter; it can be used to import the model into a different schema. For example, we developed a model using DMUSER and want to migrate it to a production schema PMUSER. You can specify this parameter as "DMUSER:PMUSER," which means the model created by DMUSER is to be imported to PMUSER schema.
tablespace_remap	An optional parameter; it can be used to import the model into a different tablespace. By default, models are exported and imported within the same tablespace. As in the case of schema_remap, here you can specify values as "TABS01:TABS02," which means the table space TABS01 has to be changed to TABS02 while importing.

You can use the following code snippet to import the models into a dump file.

```
BEGIN
DBMS_DATA_MINING.IMPORT_MODEL('CLAS_DT_1_1401.dmp', 'dmdir');
END;
```

Renaming Models

Sometimes it is needed to rename the model as per business standards prior moving to the production instance. You can do so by using the RENAME_MODEL procedure of the DBMS_DATA_MINING package.

Code Snippet

```
DBMS_DATA_MINING.RENAME_MODEL('DEMO_CLASS_DT,'CUST_CHURN_MODEL');
```

Drop Models

You can use the DROP_MODEL procedure of the DBMS_DATA_MINING package to drop a model. It is necessary in cases where we want to drop an existing model and import the new model. Let's say, we already have an existing customer churn model up and running in the production instance. We have created a new model with a lot more features and logic. If we want to have only one customer churn model with the same model name, we need to drop the existing one and import the new model that we prepared.

Code Snippet

```
DBMS_DATA_MINING.DROP_MODEL('CHURN_MODEL');
```

Predictive Model Markup Language (PMML)

PMML is an XML-based format to exchange predictive models across platforms, which is developed by the Data Mining Group. It defines the de facto standard for defining predictive models that helps to move the models between different tools and applications without the need to recreate the model. A model can be developed in one application and can be used in another application using PMML. The latest PMML version is 4.3.

PMML follows a structure, a set of predefined elements, and attributes of all the elements needed to define a predictive model. The following is a high-level overview of a PMML structure:

1. **Header**: It contains general information regarding a model such as version, timestamp, and environment information

2. **Data dictionary**: It has the information related to the data types of the field attributes (predictors and target) used in the model.

3. **Data transformation**: This section has the details related to the data transformations such as data normalization, mapping, and discretization.

4. **Model**: This section has the model details such as the model name, the name of the target variable in the case of regression and classification models, the list of predictors, and the coefficients.

The PMML code for different algorithms can have more or less information, but the structure is somewhat similar to that I discussed in the preceding section.

```xml
<?xml version="1.0" ?>
<PMML version="3.1" xmlns="http://www.dmg.org/PMML-3_1" xmlns:xsi="http://www.w3.org/2001/XMLSchema-instance">
  <Header copyright="DMG.org"/>
  <DataDictionary numberOfFields="4">
    <DataField name="age" optype="continuous" dataType="double"/>
    <DataField name="salary" optype="continuous" dataType="double"/>
    <DataField name="car_location" optype="categorical" dataType="string">
      <Value value="carpark"/>
      <Value value="street"/>
    </DataField>
    <DataField name="number_of_claims" optype="continuous" dataType="integer"/>
  </DataDictionary>
  <RegressionModel
    modelName="Sample for linear regression"
    functionName="regression"
    algorithmName="linearRegression"
    targetFieldName="number_of_claims">
    <MiningSchema>
      <MiningField name="age"/>
      <MiningField name="salary"/>
      <MiningField name="car_location"/>
      <MiningField name="number_of_claims" usageType="predicted"/>
    </MiningSchema>
    <RegressionTable intercept="132.37">
      <NumericPredictor name="age"
                        exponent="1" coefficient="7.1"/>
      <NumericPredictor name="salary"
                        exponent="1" coefficient="0.01"/>
      <CategoricalPredictor name="car_location"
                        value="carpark" coefficient="41.1"/>
      <CategoricalPredictor name="car_location"
                        value="street" coefficient="325.03"/>
    </RegressionTable>
  </RegressionModel>
</PMML>
```

Figure 8-14. *Sample linear regression PMML file*

Importing and Exporting PMML models into Oracle Advanced Analytics

Oracle Advanced Analytics supports PMML version 3.1. We can only import PMML code for regression algorithms into Oracle Advanced Analytics. Also, the PMML export option from Oracle Advanced Analytics is limited only to decision trees.

We can use the IMPORT_MODEL option of the DBMS_DATA_MINING package to import PMML code into data mining tables. The following example imports a sample linear regression PMML code to the Oracle database (see also Figure 8-14).

Sample PMML import code snippet

```
BEGIN
DBMS_DATA_MINING.IMPORT_MODEL ('LinearRegression',
                              XMLType(bfilename ('DMDIR',
                              'linearRegression.xml'),
                              nls_charset_id ('AL32UTF8') )) ;
END;
```

The get_model_details_xml of dbms_data_mining can be used to export PMML for decision tree models from Oracle Advanced Analytics (see Figure 8-15).

Sample code snippet to generate PMML for decision tree

```
SELECT  dbms_data_mining.get_model_details_xml('demo_class_dt')  AS DT_
DETAILS from dual;
```

You can save the PMML file in your disk and can export this model to other data science tools such as R or SAS.

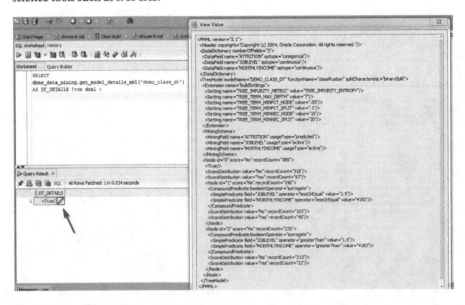

Figure 8-15. PMML file for decision trees

Summary

My intention in this chapter was to walk you through the different migration and deployment methods that can be leveraged in Oracle Advanced Analytics. You learned different techniques that are available in Oracle Advanced Analytics to move the data science models between environments quickly and efficiently. You also learned about the PMML technique to import and export the models that are created using different tools such as SAS, R, and KNIME (Konstanz Information Miner) to and from the Oracle Advanced Analytics platform.

This brings an end to our discussion on the two powerful underlying technologies of Oracle Advanced Analytics technology stack: Oracle Data Miner and Oracle R Enterprise. As the world is changing fast, so do the technologies. Oracle is continuously improving these products, and new features are added to the technology stack in every release. To keep yourself updated on these technologies, you can follow the Advanced Analytics section in the Oracle Blog as well as the Oracle Technology Network.

Index

© Sibanjan Das 2016
S. Das, *Data Science Using Oracle Data Miner and Oracle R Enterprise*,
DOI 10.1007/978-1-4842-2614-8

285

Get the eBook for only $4.99!

Why limit yourself?

Now you can take the weightless companion with you wherever you go and access your content on your PC, phone, tablet, or reader.

Since you've purchased this print book, we are happy to offer you the eBook for just $4.99.

Convenient and fully searchable, the PDF version enables you to easily find and copy code—or perform examples by quickly toggling between instructions and applications.

To learn more, go to http://www.apress.com/us/shop/companion or contact support@apress.com.

Printed in the United States
By Bookmasters